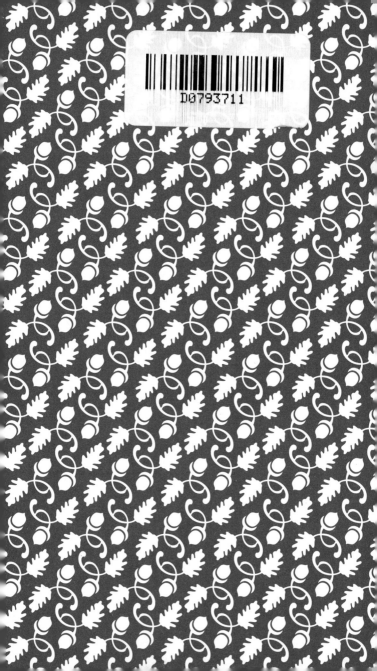

Technology and Sustainability

Withdrawn from Collection

technology &
SUSTAINABILITY

PETER DENTON

RMB

Copyright © 2014 Peter Denton

All rights reserved. No part of this publication may be reproduced, stored in a
retrieval system, or transmitted in any form or by any means – electronic, mechanical,
audio recording, or otherwise – without the written permission of the publisher or a
photocopying licence from Access Copyright, Toronto, Canada.

Rocky Mountain Books
www.rmbooks.com

Library and Archives Canada Cataloguing in Publication

Denton, Peter Harvey, 1959-, author
    Technology and sustainability / Peter Denton.

Includes bibliographical references.
Issued in print and electronic formats.
ISBN 978-1-77160-039-2 (bound).—ISBN 978-1-77160-040-8 (html).—
ISBN 978-1-77160-041-5 (pdf)

    1. Technology—Social aspects. 2. Technology—Environmental aspects. 3.
Sustainable living. I. Title.

T14.5.D49 2014   303.48'3   C2014-904033-4
                                           C2014-904034-2

Rocky Mountain Books acknowledges the financial support for its publishing
program from the Government of Canada through the Canada Book Fund (CBF) and
the Canada Council for the Arts, and from the province of British Columbia through
the British Columbia Arts Council and the Book Publishing Tax Credit.

| Canadian Heritage | Patrimoine canadien | | Canada Council for the Arts | Conseil des Arts du Canada |

BRITISH COLUMBIA
ARTS COUNCIL
Supported by the Province of British Columbia

The interior pages of this book have been produced on 100% post-consumer recycled
paper, processed chlorine free and printed with vegetable-based dyes.
Printed in Canada

MIX
Paper from
responsible sources
FSC
www.fsc.org   FSC® C016245

Contents

Preface

This book deals with choices – yours and mine. Not the choices we made yesterday, not the ones we should make tomorrow, but the choices we are making today – just as we do every day of the week.

Ordinary, everyday choices have made the world what it is right now. What we choose today will help create the world we will encounter tomorrow.

We don't seem to recognize that these rather obvious statements are powerful expressions of responsibility and opportunity on a scale that literally shapes the planet on which we (and our children) are going to live.

This book relates technology and sustainability by arguing that sustainability is only possible if we make better choices about technology today than we did yesterday.

It places responsibility for making those better

choices on each of us, wherever we live, however old we are, whatever our circumstances.

It calls us to consider what we believe is most important, what we think means the most to us personally. It calls us to understand how our values are embedded in the choices we make, every day.

This is not just another exhortation to "intentional living," however. It is recalling the truth that we live in a universe of relations, in which everything we do has an effect on other people and on the place where we live. If our global society is unsustainable – and I am convinced it is – it is unsustainable because of the poor choices individuals have made and continue to make. Every day.

This book assumes we do not intentionally make poor choices. It is normal to want the best possible things for ourselves, our children and our world. Making self-destructive decisions is as much a sign of ill health in a society as it is in an individual. Social suicide is no more acceptable than jumping off a bridge.

We have other options, we have other choices, but we need to understand who we are and where we are – not only in the trajectory of human history, but in the story of the Earth itself. If life is a

Gift that we experience within the universe of relations, then what we do with that Gift is worth some thoughtful consideration.

❦

Technology and Sustainability: Choosing the Future is in some respects the prequel to *Gift Ecology: Reimagining a Sustainable World* (Rocky Mountain Books, 2012). The core of the argument of this book is to be found in the prelude to *Gift Ecology*.

While the books may be read separately, they deal with different facets of the same problem – how to understand our relations with each other and with the Earth in order to make better choices toward a sustainable future. *Gift Ecology* looked at the historical topography of the problems we face, identifying the ideas and attitudes that underpin our currently unsustainable global culture.

Technology and Sustainability looks at the practical choices necessary to change the game into one that we have a chance of winning. It considers not only what we need to do, but how – and where – to start.

The main ideas offered in both books emerged in courses I taught at the University of Winnipeg and for the Royal Military College of Canada. After I was asked to create a cross-college course on ethics and sustainability at Red River College in 2004, I had the opportunity to develop these ideas into their current form. I am grateful to the 3,500 or so students who have wrestled with the problems of technology, ethics and sustainability over the past ten years, challenging me to refine the argument and justify my conclusions.

The gift of relationships takes many forms and wears many faces. Over the years leading to this book, I have been humbled by the support of more people than can safely be mentioned here. For their encouragement, criticism and inspiration, however, I am especially grateful to Kaitlin Bardswich, Dave Bashow, Sanya Beharry, Amna Bihery, Meena Bilgi, Michael Blair, Bob Carmichael, Cora Chojko-Bolec, James Benjamin Cole, Maggie Comstock, John Craynon, Dorothy Davidson, Kathy Davis, Susan Deane, Brock Dickinson, Fred Doern, John Ehrenfeld, Cordt Euler, Drew Evans, Michael Farris, Gerald Farthing, Bill Fraser, Hilary French, Elisabeth Guilbaud-Cox, David Giuliano, Adele Halliday, Matt Henderson, Mike Hennessy, Ray Hoemsen, Valerie Howat, Calvin

James, Beth Jennings, Calestous Juma, Alexander Juras, Joshua Ole Kanunka, Jim Kenny, Amanda Le Rougetel, January Luczak, Don MacDonald, Bill MacLean, James Marles, Carl Matheson, Jose De Mesa, Christina McDonald, Bill McKibben, Laurie Morris, Monica Mutira, Steve Nagy, Fatou Ndoye, Bill Noakes, Samantha Nutt, Akinyi Nyangoma, Maureen Olafson, Gary Paterson, Arlene Petkau, Bill Phipps, Sal Restivo, Margaret Riffell, Gordon Robinson, Nora Sanders, Diallo Shabazz, Vandana Shiva, Claudius Soodeen, Jean Sourisseau, Rob Spewak, Joanne Summers, Patti Talbot, Sylvain Therriault, Mardi Tindal, Roberto De Vogli, Dale Watts, Ken Webb, Bob Willard, Robert Young and Laetitia Zobel.

That *Gift Ecology* bridged the generation gap and was read by both my children and my parents was another gift, as was the support behind the scenes of my partner, Mona. The possibility my grandmother will hold this next book and appreciate its argument as she celebrates her 106th birthday reminds me it is never too late to join the struggle – and it is always too early to quit.

To Don Gorman and the dedicated staff at RMB, words beyond a simple and heartfelt "thank you" escape me. I am honoured to have my work included in the Manifesto series again. You make sure every word matters and every tree counts.

This book is dedicated to my father, Tom Denton – recognizing that in helping refugees come to Canada to start a new life (as he has done now for thirty years), he has always tried to change the world one person, one choice, at a time.

The Argument

Take a snapshot of what is happening around you right now and I suspect that, with changes to the places and dates, it is little different than what my own snapshot would look like as these words appear on the page in front of you.

The world in 2013 saw a succession of extreme or unusual weather events in the form of snow, rain, drought, floods, tornados, hurricanes and typhoons. Whether it is water, wind or fire – or the earth moving underfoot – the elements themselves seem out of balance. Even when equilibrium appears to be restored, if we heed the evidence of vanishing glaciers and thinning Arctic sea ice, everything seems a little warmer.

From what our instruments tell us, it actually *is* warmer – although global warming is far from the only problem that confronts us. When we add together extreme weather, environmental systems degradation, population growth, regional conflicts, refugees and internally displaced people, species extinction, lack of available clean water, housing crises, urban sprawl, energy prices, wildfires, greenhouse gas emissions, air pollution, famines and food security problems, salinization and desertification, and epidemic disease, we are brewing the perfect global storm.

Talk of recognizing and respecting planetary boundaries is met with incomprehension or denial. We don't know what this means, and if even if we have an idea, we don't know what to do about many of the dilemmas we face.

Denial is catastrophic, but so is despair. This book comes out of the desire to motivate people toward hopeful, positive and practical actions – while seeing the world for what it is and ourselves for what we are able to do.

It is a matter of attitude, of a different perspective that enables such a constructive response. This attitude is not the result of clever scientific or technological solutions – though we need more of both. Nor is it the result of some new economic analysis – though real cost accounting of our situation and our alternatives is crucial. Nor is it the result of either more education or further research, because neither will create a sustainable future in time.

And time is an issue – we don't have a generation to think about it anymore. That would have been the case for my parents' generation. We don't have a generation in which to enact the changes – that would have been my generation. Instead, the current generation – my children – will inherit the nightmare unless we do something now.

We have less time than we want, less than we think, less than we would ever have believed, to shift the planet in new directions. Archimedes said that with a lever long enough, he could shift the world. We need to find that lever, but unfortunately we make the mistake of vainly searching for the right tool, the right piece of technology, to accomplish the impossible.

What we need instead is the right story, which has always been enough to change the world, sometimes (and certainly in geological terms) almost overnight.

We need to understand the story of technology and why the story of technology is really our story, the human story, about the choices we have made. The story we need to shift the world is a moral narrative, found at the ethical nexus of technology and sustainability – how we choose and what we choose about the world we create, in which we then live.

This book approaches sustainability as a problem to be solved, not a situation to be lamented. A sustainable future is not a chimera; it is as possible as any of the things humans have accomplished in our lifetime that would once have been unthinkable outside of science fiction.

You might think understanding all the parts is the first step to solving any problem. The first dilemma we face, however, is that sustainability is a system problem and not something that can be broken down into parts. The second dilemma is that it is a dynamic system problem, whose components are constantly moving and changing.

At every point in time and in every place, something different needs to be done. We are dealing with a web of interrelationships. Some are large and obvious, but most are small and virtually undetectable. Thus, the mechanical, linear logic that the Machine Age spawned simply will not create a solution that will work anywhere for long, if it works at all. Our reactions to any situation will be too crude, too late and too ineffectual to change the underlying dynamic at the root of all the individual problems that emerge.

We need to find another way.

⚜

We are living in a global society that the planet cannot sustain. Technology is in some way responsible for this lack of sustainability.

You can find versions of these two statements in many places – and responses and rebuttals in many others.

The point remains that sustainability and technology are intertwined. What we understand about the one affects our understanding of the other. Thus we need to consider technology and sustainability together if we are to understand the situation in which our generation finds itself today.

Technology and sustainability are linked by the choices we make, individually and collectively. What we "choose" in terms of technology affects the sustainability of our lifestyle as individuals and the longevity of the society to which we belong.

In other words, they are linked by ethics. If morality is what people believe about good and bad, ethics is what they choose to do about it.

So we need to begin by understanding what lies behind our choices. Why do good people make bad choices? And what do they make bad choices about?

In answer to the first, I could fall back on a religious response. Pick your religious tradition, and you could conclude that good people make bad choices because we live in a fallen world; or because we have an inherently sinful nature; or because we live in a world of illusion in which nothing, including ourselves, is what it seems.

No doubt many millions of people believe one version or another of these statements, but in the context of choosing a sustainable future, they are all not good enough. Like playing Alice in theological Wonderland, these responses send us down a hermeneutical rabbit hole into a place where our view of the world is distorted and our interpretation of what is going on around us is skewed.

Each of the major religious traditions needs to find its own answers to the problems of personal meaning in the cosmos – and then recognize that, whatever their conclusions, to live together into a sustainable future requires a consensus about what then we all should do.

Setting aside the religious response and looking at the problem from a pragmatic standpoint, good people make bad choices because they don't think enough (or at all) about what they are choosing.

I have puzzled over this for some time: how can educated, compassionate, intelligent, thoughtful and kind people make consistently foolish and destructive choices with horrific consequences? It doesn't make any sense – unless they are not thinking enough, or at all, about the decisions they are making until it is too late.

When I talk to groups about technology and sustainability, I ask (as an opening question) how many of them have made an ethical choice so far today. Put yourself in that group, and consider that I am asking you that same question right now.

Usually, there is little response. It's that awkward opening moment a comedian has when no one responds to a joke. With some urging, one or two hands might go up, with a hesitant response. The best was from the young man who said he must have made an ethical choice that day, but he did not know what it was!

There are two reasons we live in an unsustainable global society and this is the first: we have

7

all made hundreds of ethical choices to this point in our day – all of us. We just don't tend to see or think of them in that way. So we drive cars, take buses, trains, walk, ride bicycles – however we get around, we embody in our travelling the ethical choices we make about transportation.

What we eat, how much we eat, where it comes from, whether we grow it or cook it – all these things are ethical choices about the food we consume. What we drink – even something as simple as water – reflects those ethical choices too. Whether we drink water or strain it through our coffee machines or buy it en route to our morning destination, whether it is bottled and bought or poured out of municipal water systems, reflects ethical decisions that we (and others around us) make every day, all the time.

Where were the clothes made that you are wearing right now? Do you know? Can you find out? Who made them? Under what conditions? Where were they purchased – and how were the workers treated there, along the way to the cash register where you handed over the money to buy these particular clothes instead of something else? Was the price you paid the most important consideration?

If you knew more about the conditions under which the garment was made, would you disregard the cheaper item in favour of the more sustainable one?

Half the explanation for the predicament in which we find ourselves, therefore, is that we make ethical choices, every day, all the time, without recognizing them for what they are.

Suppose we are all brilliantly ethical people, making wonderful choices at every point in our day. What chance do we have of making those same choices tomorrow, if we did not realize we were making them today? We can only repeat those good choices by accident tomorrow, because we didn't recognize them today for what they were.

Multiply the possible options by the number of people who are choosing, and very quickly you can understand the problem. If more wrong choices than right choices are possible, we multiply the catastrophic effects of this collectively accidental decision-making into the daily ecological disaster that it so often is.

We live in a globally unsustainable society because people do not think about the choices they make, every day. If they did think about those choices, or thought more clearly about what they were choosing, then all their assets – intelligence, education, wisdom – would be part of the process and would lead to better and more sustainable decisions.

And this is entirely possible. We go places on time, we show up to work or school, we wear clothes, we eat food, we obey traffic lights, we hold doors open for strangers instead of hitting them – we are surrounded by a cloud of the largely unconscious ethical choices we make every day. It is the invisible glue that holds our society together at local and at larger levels.

These daily choices are necessary and inevitable, but we can no longer afford to make good choices only by accident. We should be able to give reasons for our decisions and thus be responsible for the choices we make.

But nothing around us seems to change – or at least to change quickly enough to make a difference. We are experiencing a kind of social paralysis in response to complexity. We offload choices onto institutions (governments, corporations), but the evidence is overwhelming that it has been a colossal mistake to let "them" decide for us.

In other words, this is not only a problem with individual ethics – it is also a problem with ethics compounded into social and cultural systems. We don't feel we are responsible for all of our choices, because we have passed over that responsibility, consciously or unconsciously, to the larger groups or institutions within our society.

Our response to complexity has been to create the social equivalent of the graphical user interface introduced in the early days of computers, allowing people to click on screen icons instead of having to write their own code. We are focused on what technology does for us, not on how it works, or whether we need it.

Thus accidental ethics is only half of our problem. The other half is that we are making these unconscious choices about something we don't really understand – technology. Despite living in what everyone likes to call "the technological society," we often don't understand the nature of the technology we choose and use.

Why then do we live in an unsustainable society? Because we make choices without thinking about technology we don't understand. It is thus no surprise that things have deteriorated to the point they have in our generation. It is also clear that unless we deal with both parts of the problem, nothing much will change and the situation will only get worse. We can't do ethics by accident anymore. And if we want to make good choices about technology, we first need to know what it is.

The good news is that we can solve both parts of the problem. We can relate technology and sustainability in ways that make possible a better future than the one that otherwise will surely arrive, sooner than we think.

This book is broken down into two parts:

The first section develops and explores the idea that technology is in our heads, not in our hands, and what this means in terms of the choices we make about it.

The second section explores how values are embedded in the choices we make about technology and argues that rethinking those values is the only practical path to a sustainable future.

These two sections are introduced, punctuated and concluded with reflections on weaving a story to change the world.

Prelude
Stories Around the Cultural Fire

Every culture, time out of mind, has had storytellers. There was always some variation of sitting together around the fire, time spent gathering in community.

In the silence, as the logs sparked into the darkness, stories were told and retold, generation after generation.

This was the only way knowledge could be shared with the whole community. Specific knowledge was passed along from person to person, from master to apprentice.

But the community itself was shaped, managed and directed by stories shared in common. Whether they were of lessons learned or events explained, everyone together heard the same story. Whatever the community's individual members did with their lives, what their lives meant together was expressed in common story.

In community, stories were words of power.

There had to be both storytellers and story makers. To be a story maker was to have a rare and special gift.

Once crafted, stories may be learned, embellished and retold, but the maker – the bard – had power like none other in an oral culture.

I recall learning about the bards of ancient Ireland, held in higher regard than any except the kings – though even the kings would court the favour of the bard, who would record their accomplishments in word and song.

Their power was the power of memorial, of putting into words shared across the span of time what someone should be remembered for being or doing.

The capacity to remember and recite from memory hundreds of lines of verse is a gift few have today. It is only possible in an oral culture, in which memory is the sole record.

Runes in Ogham script could never match the bard in full voice, and so the history of the people – and what it meant – was found in story.

Out of that history came the possibilities for tomorrow – created, shaped and given voice by stories recalled and retold to the community around the cultural fire.

❧

The power of story today has arguably been watered down by the sheer volume of words swirling around us (whether stored on electronic devices or on paper) – a daunting reality that simply overwhelms more than feeble attempts at remembering.

Our communal fires have been reduced to occasional barbecues. Even communal activities are now organized thematically, not geographically – hockey tournaments and bingos attract participants but not the community as a whole.

Yet the main elements of communal narrative persist in the few television programs that generate huge audiences or the reality shows in which people become the public authors of their own stories. Nothing draws more audience participation than the tales of rags-to-riches, unknown-to-Idol, that have become the serialized mythology of our day.

To be sure, there are information shows that draw folks interested in witnessing the craft and skill of a master, yet it is an audience of vicarious apprentices, unlikely to replicate those amazing feats in their own kitchen or workshop.

One could argue it has been 90 years since the community campfire has had any real place in the lives of most North Americans, except perhaps while on vacation. But that feeling of sitting around the cultural fire has transitioned to the family sitting in front of the radio, to the family sitting in front of the television, and now to individuals sitting by themselves in front of a computer yet connected to people around the world through social media.

Around the early cultural fire, you had to sit close to the speaker to hear what was said and perhaps to join the discussion. Now geographical distance is irrelevant, reduced to the inconvenience of different time zones, as we sit around a global fire listening to the stories of our day.

Yet what catches our interest and attention today is the same as it has been from the very beginning.

We want good stories, but they need to be moral ones. There should be good and bad – and a space somewhere in between, where the action takes place.

These stories are not merely recitations. They are moral narratives. And the characters become as real – if not more real – than anyone around us.

We are not only characters in the stories of other people but also actors in the stories we ourselves create. What those characters say and do is important, however small a part they play in the larger drama, and shapes the outcome of that moral narrative.

Shakespeare was a master storyteller because he created characters that had lives of their own, expressing who they were in a few lines of dialogue, yet able to capture the imagination long after the culture that spawned them was forgotten. The moral narratives he crafted, however, played on the expectations of the audience. A beggar was rarely a beggar, and a king was rarely a king – and the dramatic tension that resulted from the conflict between interior and exterior, between nature and behaviour, drove the narrative to its conclusion.

If we want to weave a story to change our world, we need to understand that same dramatic tension, as events unfold onstage around us, as well as the role we have been cast to play – or have chosen to play – ourselves.

We need a good story. Words of power within a moral narrative still shape the spaces of our lives. They determine the trajectory of what we are able to choose.

Sustainability unfortunately fails the story test, at least at the moment. It has yet to produce a counter-narrative to Progress and Abundance.

There is no moral tale to replace that of the rags-to-riches Horatio Alger hero, boot-strapped to success, which shows how hard work eventually rewards the individual with everything his heart desires.

There is also no popular alternative to the lucky lottery winner, the person otherwise impoverished for life who spends grocery money on an impossible dream that comes true.

These are tales of want, of need, of desperation, but the moral is one of hope and possibility, rewarded not by sufficiency but by excess. Over the last century, they have also degenerated from narratives of self-sufficiency, rewarded by fate, into icons of serendipity.

We need a new sustainability narrative, one that more convincingly embodies hope and possibility but does not offer excess as the reward, one that frames the pleasant consequences of intention and does not merely record the outcome of cosmic accident.

Surveying bookstores these days, one gets the impression that current moral narratives tend to provide escape into fantasy, the ambivalent duality of good and evil in which balance is fleeting and evil is poised to prevail.

Popular stories for children often feature zombies, supernatural demons, werewolves and vampires. There is dark power, destructive but alluring, creating a moral narrative in which the good die young but the evil live forever and mere mortals are helpless.

There is much less of C.S. Lewis and more of Tim Burton, less Narnia and more nightmare. Comic books full of heroes come to life onscreen, but they never have a happy ending, just a pregnant pause before the next calamity ensues.

Thus popular moral stories abound but the narratives of anxiety and need overshadow those of generosity and gift. This is a problem, because stories mobilize us to live toward the possibilities they contain.

We are characters in the drama of the days of our lives, but we are also authors of the story in which we are cast. A new story therefore needs to be written by each of us. We need to find an opportunity to share our stories and to hear those of others, in order to recreate the local community on which any sustainable future must be based.

We can create communities of practice through the Internet, which in its own way has also become a vehicle for sharing our stories. Facebook, Twitter and other social media avenues – like blogs – allow us to share our stories around the world. But the difference here is that the act of storytelling is only part of what used to happen around the cultural fire.

Many people now tell stories – perhaps more than ever before – but they are told into the void, where there is no assurance that anyone anywhere will hear what we have to say, much less understand it. Most of what is Facebooked is only "liked" – notice there is no button for "dislike"? – without comment. We can say more, but these comments are short by convention (who reads a long post?) or are limited by the medium itself to a certain number of characters.

Anyone with more than a handful of friends or followers knows that the thoughts and important moments shared around this electronic cultural fire disappear into a maelstrom of competing stories. There is no equivalent to the convention of the aboriginal talking stick, which compelled the group around the circle to listen to the one who had the stick. Instead, everyone talks at once. No one catches more than fragments of the stories that are shared on the news feed and then disappear forever, preserved in electronic memory that records but does not understand.

❧

The story we write or retell determines the parts available for us to play. How the story unfolds shapes how we are able to act them out.

There is a paradox here when it comes to any narrative of a sustainable future. We are bombarded daily with a narrative of individual helplessness in the face of mass culture, while at the same time enticed by a narrative of personal importance, the hyper-world of "me" that focuses the compulsion of consumption on the seemingly "free" individual.

Thus we are told we can do nothing against "the Government," against "Big Business," against "the Church." We are told we can do nothing about climate change, species extinction, lost habitat or degraded farmland. But at the same time, we are encouraged to spend freely, to retire early, to satisfy whatever whims of personal desire are fanned by advertising. The only plausible constraint is the need to wait for the next "sale" to satisfy our latest impulse and feed the craving of the moment.

It is a strangely hollow and unsatisfying story, so it's understandable that we retreat into fantasy, where the heroes (however ambivalent their characters) are at least free to act as they choose against the forces arrayed against them, whether or not they are successful.

But for a sustainable future, we need a personal narrative of hope and possibility to counter this mass narrative of powerlessness and despair.

Yet however grim the situation, in fact or in fiction, however much our own characters are manipulated by the setting and the plot, we persist in story making and storytelling.

Stories are our cry against oblivion, the candle we light even though we know it will be snuffed out. And not just our stories, but our poems, our paintings, our art, our songs, anything that captures the personal in a way that preserves it beyond the moment of its inspiration or crafting.

The same is true for performance, because you

do not have to create the story or compose the music. Each time the story is told or the composition is played or sung, the performer crafts something special for the audience at hand.

I recall the compelling stories of the performance of Verdi's *Requiem* at Terezin, the Nazi concentration camp, to an audience and by performers soon destined for Auschwitz. Such things are triumphs of the human spirit, the real triumph of the will to life and meaning in the face of death and obliteration.

Stories inspire, warn, compel. They frame the meaning of our lives in ways we can never escape.

They return at unguarded moments, evading the conscious mind that otherwise structures and controls our thoughts as our routines structure our daily lives.

They reflect the rhythms of culture, the undercurrents of psyche, the ties to a universe of relations from which our activities try to distract us. The "to do" list never finally supersedes the "to be" list, however, no matter how much we push it to the back of our consciousness.

We have no idea what random series of our thoughts, scribbled hastily, might be remembered, discovered between the pages of a book or stuffed inside a wall, long after we are gone. But we are compelled to story, echoes of the cultural fire embedded in the genetics of memory, and in that remembrance lies hope.

PART I

Of Axe-Heads and Axioms

Chapter 1
Grok and the Rock

Technology is in your head, not in your hands

Technology is everywhere. Our lives are mediated in every aspect by devices of one kind or another. The Nature we encounter is harnessed, shaped and controlled, serving not only our intentions but our whims.

The discussion of ecological footprints assumes there will always be one. It is just a question of how large it will be. Yet Aborigines in Australia lived in the same place for tens of thousands of years, leaving only paintings on rocks walls as evidence of their existence. We can't manage to go camping for a weekend without irrevocably altering the landscape in some way.

The whole world teeters on the edge of the built environment – and if you factor in climate change, it already is affected, in every corner of the planet, by the consequences of human activity.

Technology as gadget is tangible, the cell phone in our hands that tells us – perhaps in a voice of its own – how to find our destination, where to shop, or the best restaurant en route, in addition to the messages it transmits and receives.

Our cars now want to park themselves, automating skill and short-circuiting the judgment that experience creates.

The process of technological creation has been foreshortened to its end product – the frozen and then microwaved food plate that replaces growing and cooking.

There are axioms of contemporary technology, principles that are accepted without evidence or demonstration. So it is understood that newer is better, that faster is better, that distance is irrelevant, that the machine will outperform anything nature can offer. Unlike our ancestors, we designate ourselves as inhabitants of "the technological society," as though our inventions and innovations have created something that has never existed before, that somehow our civilization is the pinnacle of human evolution.

Yet it is this axiomatic, unchallenged acceptance of the nature of technology and its implications that underpins the unsustainable society in which we now live and thus poses the greatest threat to a sustainable future.

Technology is not new; it is as old as thinking humans. I would argue that technology is what *makes* us human. We should be labelled *Homo faber*, "Man the Maker."

To unpack this idea, we need to go back in time to the beginning of human society and the story of Grok and the Rock:

Imagine a time before Technology, when humans lived essentially an animal existence, foraging for food, without fire, shelter, even without clothing. It was a dangerous existence; these early humans lacked the speed and the strength to defend against predators.

So it was for Grok – who was not even a caveman, because there were no caves for shelter. Every day, as the tribe foraged for food, the local carnivore (let's say a sabre-toothed tiger) would come by and catch one of the tribe off-guard and eat that person for lunch.

One day, it was Grok's turn to be surprised.

Backed against a rock wall, cowering at its base, he closed his hand around a fist-sized rock as the tiger moved in for the kill. Without thinking, he threw it straight at the tiger's face.

It was a toss-up who was more surprised, Grok or the tiger. The tiger had never experienced this kind of pain before and, startled by the unexpected, it ran away. As other tribespeople emerged from the bushes, Grok found himself the centre of attention. A normally unremarkable male member of the tribe, accustomed to getting leftovers, he became an instant hero.

No one had ever before driven off the tiger and escaped unscathed. How had he managed to do this? Thoughtfully, he picked up the rock and showed it to everyone. Somehow, it was this special, magical rock that had saved his life and scared away the beast.

That night, the female members of the tribe were particularly attentive to Grok, who thoroughly enjoyed his newfound status. The other males, accustomed to their usual rank, were understandably miffed. One man in particular got thinking – what did Grok have that he didn't? The answer was obvious: the magic rock.

That night, while Grok was sleeping, he stole the rock. The next morning, with the tribe less alert than usual because their fears of the tiger had been blunted, the tiger arrived again for lunch. He was hungry, his nose was sore and he roared to terrify his prey.

The tribe turned to Grok and cried for help. Grok staggered to his feet and looked for his rock. Aghast, he realized it was missing.

As the tribe tried to scatter, each hoping to escape the inevitable, the one who had stolen the rock swaggered forward and gestured: "Leave it to me!"

Holding out the special rock at arm's length, he bravely advanced toward the beast, shouting the equivalent of: "Go away, tiger!"

He, of course, became lunch. The tiger, puzzled at such strange behaviour, spat out the rock, ate one more person for dessert and stalked off.

By this point, the males had done some thinking – or perhaps it was the females who figured out what was going on. The rock was not special – there was nothing different about it than any of the other rocks in the vicinity. Grok was questioned directly about what he had done and the prehistoric version of the light bulb was lit.

The next day, when the sabre-toothed tiger arrived for lunch, he was met by an entire tribe of rock-throwing humans – and this time he WAS lunch.

Everything changed in that moment. In a very short time, humans not only spread over the whole planet, they hunted to extinction many large prehistoric species with tools not much more sophisticated than that first rock.

It was still just a rock. But attach a piece of wood, and it became a hammer, an axe, a club. Find two differently shaped stones, and the wild grains could be ground into something more edible and therefore became worth domesticating. Build a circle of rocks and fire could be contained for warmth and for cooking. Rocks could be used to build structures, divert water – bang two rocks together, and the rhythmic sounds of primitive music even became possible (perhaps the first rock band?).

But for which of these uses was the rock actually intended? Left alone on the ground, after all, it was still just a rock.

The rock was all of these things and more besides. Any human's choice determined the kind of technological system in which the rock would be used – and why.

What mattered in those early examples of technology was not its "discovery," but the intention that lay behind what it was then used to do.

Technology was in their heads, not in their hands. It has been the same ever since – we have just lost sight of the implications of what this means for our own society.

Our most important piece of technology – our most important tool – is between our ears, just as it has always been.

*§

To move from narrative to definition, technology is instrumental knowledge and its practice. Technology is knowledge that we use to do something. It is not new; it has been around since the very beginning.

Of course, technology is not just an object; it is also the web of interactions that surrounds the tools we make, choose and use. The idea that technology refers solely to objects hides from view many of the choices we make about it.

Nor does instrumental knowledge just involve the immediate uses of a tool; it can be potential as well as actual knowledge of how to do some-thing.

What we do with our technology, for good or ill, results from the choices we make, because instru-mentality lies in the mind of the person who de-signs, develops and uses that technology.

In some ways, this is not a new idea. I remem-ber growing up with the original *Star Trek*, with its extrapolation of the American dream in combi-nation with the ideals of the United Nations.

Woven through the *Enterprise*'s encounters with alien civilizations was the Prime Directive that forbade the use of technology more advanced than the locals already possessed. The mere sight of this new technology and what it did was dan-gerous for beings who lacked the social and moral contexts required to use it wisely. The moment the more primitive locals saw the technology, it in-stantly became possible for them to develop it – not merely the tool, but also the systems it implied.

Chapter 2
The Nature of Technology

Technology is always expressed in systems

Obviously, to say technology is in our heads, not in our hands, is to sidestep a large number of academic discussions about the nature of technology.

A statement that is so absurdly simple therefore risks being dismissed as simply absurd.

It's the application of this statement, however, not the central idea, that is difficult. It's just as easy to argue that following the Golden Rule – treat others as one wishes to be treated – would solve many of our ethical dilemmas. After all, every major religious tradition has something like the Golden Rule at its core.

The difficulty lies in our use of capital letters – which is one way of understanding what philosophers call the process of reification. What is the difference between Technology and technology? Between Nature and nature?

For example, ask people if Technology is opposed to Nature and you will likely get a "yes," accompanied by some puzzlement as to why you would even bother to ask the question. Yet there are huge assumptions embedded in both words.

The capital letters give a concrete identity to an idea that does not otherwise exist. Technology doesn't exist or have a particular identity until we say it does – and then there it is! Similarly, Nature doesn't exist until we say she does and thus give her a particular identity (as well as gender) – and then there she is!

Technology and Nature exist in our heads, as particular ideas – or rather, as a collection of assumptions, ideas and examples. Neither concept has any other kind of basis for existence except our decision to see it in the world around us. We find it easiest to talk to other people about Technology or Nature if they share the same understanding as to what the term means. Both concepts are very much part of the culture and the language in which they are expressed.

The moment, however, that we start talking

about these concepts the way we talk about something we can kick – like that rock – problems start. Philosophers have argued since Plato about whether there is a world of universal ideas that exists outside of our thoughts, to which our minds gain access when we think. Is there an ideal of Nature that exists independent of humans thinking about her? Is there some force in the universe labelled Technology of which our tools are merely poor human examples?

All that interests me here is what you mean when you use the word and whether I share the same understanding, the same ideas, the same assumptions and the same examples. Otherwise, it might sound like we are talking about the same thing, but we really are not. That is how arguments start and nothing useful is accomplished.

So when I say Technology is in our heads, not in our hands, I am really saying Technology is not an object by itself – it is knowledge that is used to do something, knowledge that we choose to develop, perhaps to make into an object.

To put it another way, Technology does not exist apart from our choices. Technology (with the capital T) should be understood as technology, expressed in technological systems that are designed, chosen and used by people. To use it with the capital letter as though it exists separate from human knowledge and choices and thus outside a human context is not only misleading – it is simply wrong.

By turning Technology back into technology, we change how we understand it: from something outside ourselves – which many people have seen as autonomous, self-directed, even a force in the universe – back into something that is inside ourselves, over which we have the same control as humans have always had since Grok picked up that rock.

In the same way, there are many ways to understand Nature. In poetry, drama, literature, across the sciences and humanities, many have tried to capture the essence of what is meant by Nature. To see Nature as Other, to see ourselves as separate from or detached from the world in which we live, allows for a distancing, a manipulation – even the destruction – of the ecological context in which we all live. (Alternatively, indigenous traditions – along with religious and spiritual traditions – consider Nature as Mother Earth, presuming that one needs to honour and be in relationship with the person who gave us life.)

In philosophical terms, reification is a fundamental mistake in discussions about both the nature of Nature and the nature of Technology. Neither Nature nor Technology exists outside of what our minds assert, and arguments that assume they do are dangerously flawed.

But this is a book about technology and sustainability – and here I intentionally don't use capital letters – not about the philosophy of Technology or the philosophy of Nature. It is also not a book about the philosophy of Sustainability. Those are arguments for another day.

Suffice it to say that by considering technology as practice, as Ursula Franklin did in *The Real World of Technology* (1989), we can find constructive ways to understand the technological systems within which we live and choose. As Franklin says, whatever our philosophical definitions of it might be, we all live in the house Technology has built. There is no longer any place outside of it where any of us may live. We are all literally under the same roof. It is a global house, one that has resulted from the assumptions (and therefore values) that lie behind the technological choices we have made to build it in certain ways and not in others.

So while it is fruitless to debate the relations between Technology and Nature, we can fruitfully explore the relations between technological and natural (or ecological) systems. (In this context, I prefer "ecological systems" rather than environmental, biological or some other subset of systems. It is a reminder that ecological systems evoke the relations of all systems on the planet, expressed in the local circumstances being considered.)

Some, like Kevin Kelly in *What Technology Wants*, see Technology existing as a force in the universe, autonomous and having grown beyond our control as part of some natural evolutionary process. I see it much more pragmatically as the sum of the choices humans have made, as they have always made them. We are no different than our ancestors. In Kelly's view, if we are worried about what Technology wants, there is nothing much we can do except to go along for the ride that started with the Big Bang. This kind of approach leaves us helpless in the face of some universal inevitability and powerless as individuals to effect changes toward a sustainable future. It is the wrong story, ultimately a narrative not of abundance but of despair.

Instead, if we are worried about some of our technological systems, we need to fix them or design new ones that better reflect the values we, as thinking humans, should have. The problems with the technologies we use are the result of our choices – and the sooner we accept that responsibility and make better choices, the better off the world will be.

As far as our tools and our technologies are concerned, we choose the future. No one else – and no other power – does this for us. No technological system exists outside the choice or choices of the persons who created it. Even systems of artificial intelligence (like robots) have their characteristics chosen for them by humans, who set the parameters of the system, including what it can learn and how.

The options are clear and simple: either Technology is something bigger than ourselves, outside of our control, and autonomous in the face of human needs and desires; or instead Technology is technology, a product of what humans choose, reflecting their thoughts, dreams, desires and values, expressed in a web of social, cultural and ecological contexts.

Whether we choose our future or merely submit to it, therefore, depends upon our understanding of the nature of Technology/technology. Every example of technology we can identify is part of a technological system that exists only because someone decided to create it.

⁓

Because technology is instrumental knowledge and instrumentality lies in the mind of the user, technology does not exist in the abstract or in isolation from its context. Whether it is a rock used to stun game in the pattern of the hunter/gatherer society, or a scratch plough used to prepare a certain type of soil for a certain type of crop, or a particular alloy to improve conductivity in a microchip, technology is always expressed in systems.

A technological system is two or more instances of technology combined to produce an expected and repetitive result. Technology is useful if it does the same thing each time you use it. If the bow shoots the arrow straight the first time but sends it sideways the second time, it is not very useful. You want the boat to float every time, not just some of the time.

If you look at mechanical systems, the need for regularity and repetitive results is even more important. We design machines whose outputs are more and more precise, needing fewer and fewer human (and therefore unpredictable and unreliable) inputs. Consider, for example, the difference between car manufacturing in 1905, 1955 and today in terms of how much the worker has to do, and you will understand the point. From the craftsman making a one-of-a-kind machine, to an assembly line with a large number of workers doing the same job over and over again, to a line "peopled" with machines and robotic technology, we have refined the system so it does only what we want it to do, all the time.

In every culture, technological systems are interwoven with human systems, which is why one of the growing edges in contemporary engineering is "human factors engineering." Human factors engineering emerged from the realization that no matter what kind of a technological system you design, in the end it is used by humans. Human factors – everything from our anatomy to our attention span to our need for food and bathroom breaks – must be considered in the design and function of any successful technological system.

If technology is old, however, so are technological systems. While it is a fiction that old or "primitive" technology is simple (complexity is not a modern innovation), the number of elements within our human constructions has increased, creating system dynamics that require new means of control.

Natural and organic systems have evolved complex systems of coordination over thousands of years, systems that computers are attempting to mimic with limited (though, supporters would argue, increasing) success. The open question is whether we will ever understand this complexity well enough to have such control of the complexities of organic or ecological systems ourselves.

Western culture has often been called "machine culture" over the last two hundred years because predictability of operation in repeated circumstances is a necessary characteristic of Western technology. The machine has been the dominant analogy of life in Western thought for some time; only those things that can be identified as "mechanical" have a place in the "reality" either recognized or constructed by Western technology.

Mechanical operations are observable and predictable. In a mechanical system, there is no place for gods, spirits, feelings or elements of other human systems. Yet in cultures where an explanation of results rather than their predictability is most important ("why did this happen?" as opposed to "how does this work?"), technological systems might include supernatural or philosophical/religious elements.

These other systems can be just as valid and useful as Western scientific ones. While difficult perhaps for Westerners to appreciate, prayers to a deity or invocations of fertility spirits may be a crucial part of technological systems because they encourage active participation by people who could make other choices instead.

Uses of technology we would regard as silly or superstitious in our culture might be completely meaningful in another culture. No doubt future cultures will look at our technology and how we regard it, and marvel at the foolishness reflected in how we understand and manipulate the world in which we live today!

There is a technology of eating as well as of building. Try to eat strange food in a different cultural context with unfamiliar utensils. While everyone else is laughing at you, remember that technology is interwoven with the society in which we live and the culture in which we have grown up. If you were to take your dinner companions into your house, eating your food with your utensils in your way, they would all look just as silly.

Thus technology is knowledge used to do something. Tools are the means to accomplish some goal – after all, a tool without a use, even a potential use, is not a tool. Yet there is no point to analyzing a single piece of technology without understanding the systems within which it is placed and identifying the purposes inherent in these systems – the reasons why some particular tool or technological system was developed and used.

The cell phone is useless as a communication device without a cell tower; one cell tower has to be connected to a series of towers for the signals to be transmitted; someone else needs to have a cell phone, or you will have no one with whom you can carry on a conversation. In the absence of electricity, and how it is generated and stored, the cell phone wouldn't work; without the science needed to understand electromagnetism, the phone wouldn't work; without the plastics (from the petrochemical industry) and the metals (alloys and exotic materials like coltan), there would be no physical phone; without the electronics industry and microchips and circuits, your cell phone might be the size of a house. The list could expand to encompass many of the systems in our society today, found around the world just as much as in your backyard.

Of course, those systems are not only mechanical ones. Coltan comes mostly from the Democratic Republic of the Congo, where too much of it is mined and transported under conditions that would not be permitted by labour or other laws in Western countries. Petrochemical by-products like plastics are tied into the global oil market and the politics that goes with it. The cell towers needed for transmission have become a forest of metal and wires disrupting the countryside; and one of the questions still unanswered is what are the health effects – on everything from honeybees to the human brain – of the frequencies used in the phones. Our culture also has changed – no longer can you escape being contacted 24/7, because now a global positioning system can track you through your cell phone even if you don't want to answer it.

This kind of analysis of technological systems could lead you to think that they are recent rather than ancient. We should not underestimate, however, the complexity, range and longevity of earlier technological systems. Nor should we overestimate the reliability or robustness of our own systems.

If technology is what makes us human, then every culture since Grok and the rock has had its own set of technological systems, chosen for reasons that were appropriate to its time and circumstances.

✎

Chapter 3
Technology by Choice

*Technological systems are interrelated with social
and cultural systems*

When we consider the complexity of the interrelationships among technological, social and cultural
systems, some of the problems involved in making
better choices about human interactions within
ecological systems become apparent.

First, every culture in history has had its own
technology and made its own choices about what
technology to develop and use. As long as that culture continued to make appropriate technological choices, it survived; when it stopped making
appropriate choices, it disappeared. More specifi-
cally (because people, not cultures, make choices),
as long as the majority of its members made appropriate choices about technology, both those

individuals and the culture itself survived. When the majority of the members stopped making appropriate choices, like Grok's unfortunate friend, they perished – and so, eventually, did their culture.

These technological choices had to be appropriate to the circumstances in which the decisions were made. There is never only one answer, any more than there was ever only one rock. The right choice is the one that works, for the right reasons, to achieve the right end, at that particular time.

Second, recent is not necessarily "better." Interrelated systems of technology, society and culture in any civilization – past or present – are designed to fulfill the needs that its members decide at the time are more important than any others.

This has serious implications for the global hegemony asserted by the science and technology developed out of Western European culture in the past 200 years.

The balance of our self-assessment shifts when we realize that this dominance is the product of our own assertion. We have forced it upon the rest of the world by means of a succession of imperial means – first with religion, then with guns, then with colonial economics and finally by means of a supposedly "scientific" and "secular" consumer culture that undermines the strength of local knowledge and tries to eliminate any different personal choice.

While it will take a book of its own to unpack the implications of this assessment, here it is enough to say that technology is always in context, that individual choices and local decisions are inherent in the wise development and use of any technology. When the decision point is shifted elsewhere, for whatever reason, then the loss of local control by individuals over what happens in their daily lives has never had a good outcome for long.

So appropriate technologies (or technological systems) are the product of local choices, not ones made from afar and imposed upon local situations.

We can look back in history and assess the success and failure of other civilizations with respect to the technologies they chose to develop and use. We might in their experience find lessons to apply to our own.

Third, complexity, subtlety and sophistication

are not recent discoveries. Ancient civilizations had characteristics to rival or surpass any of the achievements of Western industrial culture. I remember going down into the giant burial mound in Newgrange, Ireland, dating to 3200 BCE and constructed with hand tools. The central burial chamber, a hundred feet underground, was illuminated by the sun at the winter solstice.

How this was accomplished, without the measuring skills, tools or mathematics of our generation, we will never know, but the same sort of complexity and sophistication is found throughout the ancient world, from Mayan and Incan civilizations, to Sumerian, to Egyptian and to Asian – essentially, anywhere we have been able to find and correctly interpret the remains of cultures that have since ceased to exist.

We also do not have to look to other cultures elsewhere to reach the same conclusions. Consider the triumphs of Roman engineering – roads built from local materials, with local labour, in local conditions; and the bridges and aqueducts that continue to function 2,000 years after their construction, carrying traffic of a kind and in a world far beyond the imagination of their builders.

Consider the intricate artistry of medieval cathedrals and how difficult it would be to match that artistry today. Nor would we find it easy to match the feats of mass construction embodied in the Pyramids or in the Great Wall of China.

In the event you still wish to argue for the unique complexity and sophistication of modern engineering solutions, try to imagine the dynamics of an 18th-century British man-o'-war, at sea on blockade and in battle continuously for decades, and see what luck you might have describing the mechanics of the ropework and the strains and stresses of the sea that a ship's master had to understand – much less how to manage a crew and fire the guns with precision.

Realize the worldwide industry that supported sailing ships and how that system functioned, and you will understand complexity is not new. Consider for how many centuries the sailing ship, in its various forms, was the sole means of global transportation, for everything from trade to warfare, and realize how much younger and more fragile all of our current transportation systems are by comparison.

Technological systems are always intertwined

with social and cultural systems, however. It is these social and cultural systems that determine what technologies are used and for what purpose. The sailing ship was used as a means of exploration, to find new lands for the Europeans to settle – or (more correctly) to conquer. It was a lifeline for those who had to leave Europe to find a place where they could build a community free of religious or political persecution for their children; it was also the means by which African families were torn apart and carried thousands of miles away to a place where they lived as slaves. It was a means of gathering food in the form of fish, to sustain life for many; it was a means of dealing death through warfare. A source of trade and commerce to generate wealth and prosperity, it was also the means by which the pirates or privateers could ruin others for their own benefit.

So, looking at all these uses of the sailing ship, was it a good or bad technology? Obviously, the moral questions about whether it was good or bad lead into the ethical questions about how someone chose to use it. Every technological system has embedded within it the values of the people who invented or developed it and those who chose to make or to use it. As a result, no technology is ever morally neutral, because it is the outcome of a series of choices made by individuals and by societies.

Whether the sailing ship was good or bad depends on how it was used, to what end or purpose. To say that technological systems cannot be separated from social and cultural systems, therefore, requires us to understand societies and cultures. If we want to understand the technology that they develop and use, we need to accept that invention is only a very small part of the equation.

Technological systems are thus not only about the objects of technology. They involve all of the dimensions of technological choices that lie behind the objects – the systems within which those objects are embedded, the intentions of the choosers as much as the outcomes of the users. We therefore need to consider the relationships between technological systems and society, just as we need to be mindful of the relationships between technology and culture.

Those relationships, moreover, are just as intentional as any individual choice to use an object of technology might be. They are just operating on a larger scale and at a distance from what the individual person might see as the horizon of his or her choices.

When we consider the relationship between technology and society, we look at the external characteristics, those external relations between technological and social systems.

So for example we might think of the automobile system, how cities and countries are designed to be traversed by automobile, affecting everything from urban planning to the positioning of gas stations, repair shops and traffic lights, and the rules of the road that go along with having a large percentage of the population behind the wheel. Society shaped the design and function of the automobile, just as – in turn – the automobile has shaped the society in which we live.

But we also need to consider the relationship between technology and culture – those internal characteristics, what is inside our heads that affects the technological choices we make.

In the case of the automobile, there is a "car culture" that makes getting a driver's licence, rather than gaining the right to vote, the real rite of passage from childhood to adulthood. Car culture is so pervasive and powerful that we hesitate to take away the driver's licences of those too old or otherwise incompetent to respond to the challenges of safely operating a vehicle. Balancing the need to drive against the need for others to survive, car culture arguably protects the rights of the driver against the rights of any potential victim.

Similarly, car culture undermines efforts at mass transit beyond the issues relating to population density. We go where we want to go, when we want, and resist the idea of sharing our vehicle with anyone – especially a stranger – who might interfere with those personal freedoms. So our roadways clog, our air becomes polluted, as people drive wherever they choose – by themselves, at low speeds, in cars grossly over-powered for the rate at which they are permitted by law or circumstance to travel – because they don't want to share their ride.

Thus it is not enough to unpack the technological systems about which we make choices every day. We must also unpack the social and cultural systems within which that technology is found, those systems whose influence on the technological choices we make may be both more persuasive and insidious than we would otherwise recognize.

Ultimately, both the objects of technology and the technological systems we choose to design, develop and use are embedded with the values we hold as individuals, as cultures and as societies.

The crucial implication of the idea that technology is in our heads, not our hands, is that technology is inevitably the product of our choices. We make these choices for reasons and those reasons reflect our values – those things that we consider to be important enough to affect the choices we make. To say, therefore, that technology is in our heads ultimately makes technology about the values we hold as individuals, as members of a society, and as part of the culture within which we live and make our decisions.

The story we need to create – the moral narrative on which a sustainable future depends – hinges on the values we hold as individuals and as members of our culture and our global society.

Technology isn't just about something we can hold out at arm's length and examine at a distance. Technology is also inside our heads, an integral part of who we are as individuals and as a culture.

To put this another way, we need to distinguish

between axioms and axe-heads. The unspoken – and therefore unexamined and unchallenged – ideas we have about Technology disguise or hide the choices that have been made about what kinds of tools should be developed and used.

Axioms are statements that are self-evidently true, established without the need for demonstration or proof. Yet even in mathematics, axioms need to be regarded with a healthy degree of skepticism.

There is much we can learn about the nature of Technology by examining it more closely, by considering those statements about Technology that would otherwise seem obvious and by considering its reciprocal relationship with both society and culture.

We can also learn a great deal about those axioms, those otherwise unchallenged values and assumptions, by taking a close look at the axe-heads themselves.

For example, it is helpful to use some distinctions Ursula Franklin identifies in *The Real World of Technology*. Technology in terms of making things may be divided into two main types: holistic and prescriptive.

Holistic technology (which may be understood as "craft" technology) places control of the whole process in the hands of one person, who decides what is to be made, chooses the design, selects the materials and puts the object – the widget – together. While multiple items may be made by the same person, the nature of the process means that each is unique, like a work of art.

In contrast, prescriptive technology breaks down the process of widget-making into a series of discrete and independent steps. Different people are responsible for different steps; while there may be a "manager," the manager oversees the *process* of production, not the production itself. Prescriptive technology is at the heart of any system of mass production, producing multiple widgets that are more or less the same, provided that the same process is followed.

Thus craft technology, while it produces unique items, depends entirely on the skill of the craftsman; some items may be wonderful, but others will likely be poor, as personal skill levels vary. Prescriptive technology, while it is unlikely to produce unique works of art, will also not likely produce widgets that are poor quality; the factory system led into the production of items not only in greater numbers, but of higher average quality.

Obviously, Western technology since the Industrial Revolution has focused on the development and application of prescriptive technology. While Franklin points out that prescriptive technology is hardly new (citing ancient Chinese bronze manufacture as an example), prescriptive systems have replaced craft manufacture in areas where volume of output is desired.

In addition, as systems of prescriptive manufacture have multiplied, so too have the means by which such production is controlled. Managers – those people who create and implement control technologies – have become increasingly important in systems of manufacture. As more and more manufacturing tasks have been taken over by machines, the proportion of managers to actual workers has increased dramatically across all elements of Western society.

Prescriptive technology obviously comes at a cost. Unique design is sacrificed to design elements that can be mass produced; the investment in skill development found in craft technology is replaced by investment in the process of production, so that skilled workers are replaced by workers whose primary skill is compliance with what the system requires for the completion of particular steps, in sequence. Put another way, individual creativity is replaced by obedience.

While this representation of the difference between the two approaches may seem to favour holistic technology, the reality for us in the 21st century is that without the widespread use of prescriptive technology, much of what is found in our society would simply never have been created or manufactured. As the global population increases, it is only through prescriptive systems of manufacture – whether of food, clothing or shelter – that such population can be sustained.

What we make, however, and how much of it, is another problem. Prescriptive technology has been attached to the idea of the machine since at least the early Renaissance period. As I wrote in *Gift Ecology*, the machine analogy – the attempts to mimic the movements of living things (especially humans) through the use of machinery – has led to an equally mechanical understanding of the means of production.

With such an approach, the perpetual machine

(the ideal mechanism, which was pursued as eagerly by Renaissance scientists as the Philosopher's Stone was sought by their predecessors) would produce whatever products it was designed to produce as long as the raw materials were fed into the hopper. The only limit to its production was the limit set by the availability of the raw materials. Questions about whether more of the product was needed, or whether it was a good idea to use raw materials in such a way, are external to the process itself. Once the machine starts operation, there is no decision, other than turning it off, that affects what it does.

The environmental – and perhaps social – problems faced by 21st-century society have their roots in the machine analogy, as it drives the philosophical engine of prescriptive technology. According to Henry Ford (who is often unjustly credited – or blamed – for the invention of mass production systems), mass production precedes mass consumption and alone makes it possible. In other words, if we didn't make so much stuff, there would be no need to figure out how to promote its consumption.

Because there is no limit to what the ideal

machine can produce, there is also no limit to what humans might need to consume in response. There is also no point, internal to the process, at which someone looks at the overall picture and decides whether more widgets are needed. Instead, humans comply with the system and continue to "do their part" as consumers to ensure its continued operation.

If this seems like a circle without end, it is – but the unfortunate reality is that there are limits both to the availability of raw materials required and the capacity of the planet to absorb what is produced. We do not have an infinite supply of inputs, nor can the planet cope with the increasing outputs, particularly of pollution, that such a system entails. In other words, the problem is not the prescriptive system of technology itself, but the mechanical philosophy that employs it – we are not machines, nor do we live in a mechanical world. If our technology could reflect the organic nature both of the world and its inhabitants, then it would more likely be sustainable.

Prescriptive technology becomes even more of a problem, according to Franklin, when it is applied not only to production but to society itself.

Franklin points to the need for compliance if prescriptive systems are to work – when making widgets, this makes sense, as that is the only way such a linear system of production can function. Yet when you apply prescriptive technology and its mechanical philosophy to people and the solution of social and cultural problems, something goes seriously awry.

People are not widgets; human problems should not be considered in terms of their relation to Gross Domestic Product. While economics may be a good measure of the successful manufacture and sale of widgets, it has little to do with happiness, or satisfaction, or the state of one's soul. Just as craft technology celebrates – for good or ill – the individuality of the craftsperson, prescriptive technology denies such individuality for the sake of the whole process, and the celebration of uniqueness is replaced by what Ursula Franklin called "designs for compliance."

Anyone who uses a computer is educated or coerced into compliance in order to use the technology. To log on, we must put in the right password; to run a program, we must install it in a certain way; and if we forget the updates or a myriad of related support activities, we will be pestered re-peatedly until we comply. Whether we need or want any of these things is increasingly irrelevant, as the decisions are taken by the makers of the soft-ware, and not by its users.

In terms of characterizing 21st-century techno-logical systems, therefore, we can identify a pre-dominance of prescriptive over holistic technol-ogy; an overwhelming emphasis on mechanical systems of production over organic systems of liv-ing organisms; a preference for a culture of compli-ance over a culture of creative individuality; and a denial of individual responsibility for the deci-sions about the use of types of technology that are wreaking certain social, cultural and environmen-tal havoc.

Our culture's inability on the whole both to recognize and accept the truth of these observa-tions is at the root of our incomprehension of what technology means today. We don't see how what we do as individuals, and as a society, affects the planet. We choose not to see that there are other ways of "doing business." We refuse to accept the personal costs that might come from changing how we live, and thus we deny any responsibility for the consequences of the poor choices we continue to make.

The problem (and its solution) is not what Technology wants. It is what we want and how we use technology to get it.

Interlude
Were the Luddites Right?

We can see the picture in our mind's eye: the crowd of poorly clothed workers, desperation and anger on their faces, a mob surging against the doors of the factory. Breaking through, they vent their frustration on the machinery inside, smashing it to splinters with the clubs and hammers they wield.

Then, outside, the sound of horses, the tramp of feet marching, and the soldiers arrive – breaking heads with musket butts, riding down those fleeing, and clubbing those they did not arrest, determined to crush the resistance to Progress as decreed by the authorities in control.

As much as they have become a meme of futile resistance to the progress of Machine Civilization, the Luddites are more memorable in popular story than, in fact, they were significant in the history of early 19th century British industrial society.

Yet the symbolic futility involved in smashing the weaving machines, and the fierce response of the regime that crushed those protests, both still strike a chord in us today.

So while the Luddites may be publicly ridiculed as backward, illiterate in the language of inevitable Progress and seemingly consigned to the chuckle-bin of history, privately, it seems, it's another story.

The meme persists perhaps because we admire their courage, however futile their actions were in the end. Their story is retold because we too face the kinds of major changes they faced, and what-ever our public expressions of ridicule, our sympa-thies lie with the Luddites and not with the face-less regime that utterly crushed them.

The story is retold not because we need a re-minder of the futility of resisting Progress, but because the Luddites were right. The Industrial Revolution presaged by the weaving machines did everything they feared it would do.

They were concerned that the machines would destroy not just the fabric of their communi-ties, but the communities themselves. The men would be reduced to unskilled automatons and the women and children to slaves of the Machine, tied to the rhythms not of nature that ebbed and flowed, but to the eternal and insistent rhythms of machines that never stopped. This is exactly what happened.

We rationalize this devastation by pointing out the benefits of industrialization, but however inevitable these benefits might seem to us, they were not obvious at the time. Those who suffered most received the fewest benefits of all. Those first affected by industrialization could see and taste what had been lost, but between the din of the factory and the taste of the coal dust, there was little advantage to be found.

Yet it was the story of Progress that drove these changes, just as it was the story of the railway that later drove private investors to financial ruin by building another few miles of a track to nowhere, decades before the railway system became a paying proposition. They were told the story of inevitable prosperity, but it was again an outcome not experienced by those whose fortunes were shovelled like cheap tinder into the firebox of its development or whose communities were bypassed into oblivion when the rail line was laid somewhere else instead.

There is a lesson here about sustainability, a

cautionary tale that demonstrates how the onset of industrial civilization revealed the dark side of the Machine.

When Fritz Lang sent his minions marching into the infernos of *Metropolis* (1927), it was merely the latest version of industrial hell that had already been played out many times before, worst and most recently on the battlefields of the Great War of 1914–18.

A decade later, it was Charlie Chaplin in *Modern Times* (1936) who captured both the rhythms of the Machine and its dark side, as the Tramp tried so hard to find a place somewhere within a society grounded in regulation and framed by linearity. No matter what he does, he is unable to fit, and so ends the movie walking with his sweetheart into the sunset, expressing an absurd hope for a better future that is not likely to occur.

We still remember the Luddites and their hopeless protest against a future that would lead inexorably to the Great War, the Depression and beyond into some future Machine Civilization that we fear will devastate our own lives and communities as well.

Or so the story goes.

❧

It's such a good story that we wish it actually happened.

Unfortunately, it is embedded in popular understanding even though it bears little or no resemblance to any of the facts. The story has grown in the telling into something that says more about us than it does about the Luddites or their fictional leader, Ned Ludd. We want it to be true, and so it is.

"Ned Ludd" was a useful leader – one whose name could be invoked in the same way as Robin Hood's, but who could similarly never be captured or disgraced. His "followers" were not victims of the Machine, but mechanics and workers unhappy with management and with the conditions of their employment in the years leading up to the Battle of Waterloo, after decades of war with France.

Their protest was against a social and economic system that was not providing them with wages or employment as they thought it should, not against advances in industrial technology or Progress as a whole. There were riots, and factories smashed by mobs, but to make this a protest against Machine Civilization requires a leap of imagination that does not survive contact with the evidence, as Steven E. Jones outlines in *Against Technology: From the Luddites to Neo-Luddism* (2006).

We retell the story – and the level of popular consensus about the Luddites is quite astonishing – because the story reflects our own concerns about the dark side of Technology and the futility of resistance against it.

As much as Kevin Kelly (among others less eloquent) tries to persuade us of the triumphs of evolutionary progress that have culminated in the technological society (his "technium"), psychologically we remain unconvinced. Others might sow the dream, but too often we seem to reap the nightmare.

Despite the persistent messages about inevitable progress, growth and development – to the point of social and cultural badgering that is eerily reminiscent of George Orwell's *1984* – the promises seem increasingly hollow. The examples of new technology we purchase barely outlast their unpacking from the boxes we are told to save against exchange or return (but rarely for repair).

New appliances are touted to be better (and cheaper) than their predecessors, but evidence of their unreliability mounts as the piles of unfixable machines grow. Rather than sitting back and enjoying the leisurely ride into the future promised by a Technology that knows what it wants, we work longer hours for less money, with less leisure than ever before, to acquire technologies we constantly need to update or replace. As the gap grows between the very rich and the rest of us, even if we know what we want, we also know we will never have it.

Those without work are spiralled into a hole out of which it is hard to climb, surrounded by the insatiable consumption that drives global culture. Try to live without television, internet, computer or cell phone in an urban environment. Whether or not you are actually invisible and voiceless, that's certainly how it feels.

We do not need to be convinced there is a dark side to Technology and so the mythology of resistance to the tyranny of the Machine is perpetuated alongside the mythology of inevitable Progress. Yet both only have substance if we give it to them.

The irony, of course, is that there is no such tyranny, just as Technology is neither hero nor villain. To repeat the stories of resistance merely strengthens the underlying problem of our unsustainable culture, because it deflects attention from our choices and renders us into heroic but hapless (and helpless) victims of the Other.

There is no dark side to Technology, because Technology doesn't exist as an entity outside of our choices. But there is a dark side to the people who make and enforce destructive decisions about the kinds of technological systems we develop and use.

I like a good story, so I am a fan of *Metropolis* and *Modern Times* and cheer for the Luddites even though I know what the ending will be. The Luddites were right, but however romantic their story has become in the retelling, it has ceased to be useful.

We need a new story, a new moral narrative, one that embodies resilience and regeneration instead of merely obstinate (and futile) resistance to the dynamics of social and cultural change.

We need to choose the future, not just watch it unfold.

PART 2

Of Soups and Systems

Chapter 4
Reverse Engineering

Choices about technology are made for reasons and those reasons reflect our values

Often people will talk about technology as though it is neutral or objective – it's just a rock, or a cell phone, or an automobile. Yet technology can never be "value-neutral," because it is the result of a series of choices that may be traced back to what we think is important.

Values are the beliefs the people have about what is important. They are often expressed in terms of judgments about what things are more important than others. If we don't think clearly enough about our choices, the reasons (and the values that go with them) are often hidden or disguised.

There is a lot of talk these days about the "unintended consequences" of our choices, especially about technology. The consequences might be unintended, but a realistic view of the values embedded in the technological choices can give us a good idea of the outcomes. We simply should not be surprised by much of what happens.

The choices surrounding the development and use of technology reflect the values of whoever makes the decision, which can be a society, a culture, or an individual.

There is never just one way of doing anything, because what we choose reflects our values.

This means we can reverse-engineer examples of technology – and of technological systems – to discover what values lie behind them.

We can also learn something about the decision-makers by examining the objects they have chosen to develop, to make and to use.

It's kind of like contemporary archaeology. Just the same way an archaeologist can learn about how people lived a long time ago by digging through the debris of their civilization, we can learn about our own values by examining the examples of technology around us.

At the risk of mimicking Andy Warhol by focusing on one iconic image, what could we learn about our culture – and its values – by examining a can of tomato soup?

We would learn that food is consumed at a distance in time and space from where it is grown, and so most of the people who consumed this product were not likely to be agricultural producers themselves. If they grew their own tomatoes, then they would not be interested in canned tomato soup from somewhere else.

Metals must be cheap, because the can is designed to be mass-produced and the container therefore is likely worth much less than its contents. Also, this is not a one-of-a-kind item – you can see by the regularity in its shape, and the number of other cans in the garbage heap, that there were many of these items.

The container is sturdy, able to be stacked, and therefore able to be transported over long distances and stored for long periods of time. This means there are many more consumers than agricultural producers. There is also a sophisticated transportation and distribution system to deliver the product, as well as a common means of opening up these cans for consumption. There is obviously a limit to how much tomato soup any one person can eat; the more cans there are, the larger the area over which they have to be distributed in order to sell.

The label is colourful and shows various symbols, so the culture relies on visual appeal (for advertising) and symbols for communicating information in addition to the words that are used. This, in turn, means that only marginal literacy is required to choose and use the product.

The bar code indicates computer technology is used for tracking the soup, which is useful if there are many cans, widely distributed.

The breakdown of ingredients shows interest in what the soup contains, and the percentages of daily intakes indicate a science-based understanding of nutrition – making this a "scientific culture," in other words. This culture apparently places a high value on the ratio of certain contents, although the nutrient list is very limited.

The recipes on the back reveal that only a crude cooking ability is required to use the product.

Despite the many things indicated on the label about the contents of the can, however, we have no idea how it tastes or whether that is important to anyone who buys it. Once people get used to the taste of canned soup, then real tomato soup won't taste right – the presence of monosodium glutamate and the high level of salt in canned soup create a flavour homemade soup will never have.

If we excavate the next layer of what this iconic soup can means, our systems analysis reveals other things about the choices behind it.

At the first level (call it T1) are the choices relating to design and production of the first item, the prototype. The materials needed must be chosen, both in terms of contents and packaging, so the item is designed using what is available to the designer/maker.

In this instance, metals must be sourced and turned into an appropriate can for the soup. Similarly, paper must be sourced and the machinery must be available to turn it into labels. These are by choice, not by necessity – there are other forms of packaging than metal cans and paper labels.

The contents must also be sourced, finding not only the right ingredients but determining the process by which the soup is made.

At the second level (call it T2) are the choices and values related to the issues of mass production.

We have one can of soup – how do we make thousands, even millions of them? What design changes to container, contents and the manufacturing process itself would be required to make mass production feasible?

For example, the ideal container might be a glass jar, filled by hand, so that the acidity of the contents doesn't cause the container to leach toxins into the soup. But it is hard to mass-produce items this way. Similarly, it is better to make condensed soup rather than ready-to-eat soup, because the container required is smaller and the whole process (figured at gallons per minute) allows for more cans of condensed soup to be produced. Instead of shipping it ready-to-eat, instruct the customer to add water or milk to create the final volume of soup.

You might have found a wonderful source of beautiful tomatoes – but they are wonderful because they are carefully hand-cultivated, carefully picked, and only available for one month in the year. You need many more tomatoes, year round, and so settle for much lower quality to get the volume and regular supply – which, in its turn, means adding salt and extra flavourings to disguise the lower quality of ingredients (like monosodium glutamate, for example, which can make shoe leather taste like steak!).

Too much milk requires the soup to be cooked at a lower temperature and stirred constantly to prevent scorching or burning, but less milk, more water and additional oil means it can be cooked faster with less risk, therefore producing more cooked soup per hour.

If you locate your production plant in a particular area and encourage the farmers nearby to grow your ingredients, you are less likely to be affected by market fluctuations in terms of cost or availability. Higher production volumes in season, when the prices are lowest, can work if the soup is preserved and processed for a longer shelf life.

Again, all these decisions are by choice, not by necessity. The primary values expressed here are to increase the volume and decrease the cost of production of the soup – not what it tastes like, or whether it is good for the consumer. These values at the T2 level drive changes at the T1, or product-design, level.

At the third level (call it T3) are the choices relating to the distribution of the product. There is no point to making millions of cans of soup if they can't be easily shipped to customers who will buy them. Values here around the marketing and distribution of the mass-produced product also drive changes at the T1 (product design) and T2 (production) levels.

For example, glass jars are fragile and prone to breakage unless they are made of heavy glass and well protected in cardboard boxes. Waxed cardboard containers leak, unless they are lined with something that protects against seepage or puncturing, and they need to be square to fit as many as possible in a case. Cans, on the other hand, are durable, next to indestructible; they can be handled at higher volume and speed by machinery; and they do not need additional protection for shipping.

If the distance between producer and consumer is large, due to the volume of units produced requiring a larger distribution area, then there are other advantages to cans, which may be filled, sealed and then processed at higher temperatures, thus giving them a longer shelf life as well as greater ease of shipping. (If bottles or cardboard containers need additional processing, radiation might have to be used to ensure all harmful organisms are killed.)

The location of the production facilities, in terms of easy access to major transportation hubs or large markets for the product, can reduce shipping time and costs.

Along with getting the product to market, T3 choices also relate to sales and marketing. How much can we sell it for? Should we make something high volume (which is what the system is designed to do) and low cost, or should we scale back production and raise the unit cost of our soup? Competition is inevitable – how do we persuade stores to sell our product? How do we persuade people to buy our product rather than our competitors'? What selling features do we advertise and how do we advertise them?

At a T3 level, if the customer doesn't buy the product or buy it at the price that is needed for profitability, then these things by themselves can drive major design changes at T1 and T2 levels. You may solve all the other issues for T1 and T2, but if the consumer (for example) doesn't like the idea of eating luncheon meat made from seals – or if $10 is too much to pay for a can of gourmet tomato soup – changes have to be made in order for the company to stay in the business of mass production.

One could no doubt write a book on all the system elements at the different levels of tomato soup production, marketing and distribution, but you will now understand my point. The can of soup you hold is not an accidental product, nor is it the inevitable result of some grand scheme of universal evolution. It is the result of a series of specific choices, made for reasons that reflect the values of the people who made them. More precisely, the final product reflects the hierarchy of their values – not just all of the things they considered to be important, but also the order in which those values effectively are ranked.

Thus, the imaginary soup manufacturers would react angrily to the idea that they did not care about the health of their customers, for without satisfied customers they would not sell any more soup. But they cared *more* about getting a supply of cheap tomatoes and maintaining their profit margins against competitors by keeping production costs low, so they used poorer tomatoes and chemicals to enhance flavours, chemicals that had not actually been *proven* to hurt the health of the people who ate them in small quantities, once in a while.

But beyond the values relating to the systems of

mass production (applied to soup), there are other, more subtle but equally powerful implications for both society and culture. Selling people canned soup requires us first to sell people on the *idea* of canned soup in the first place. Henry Ford wrote, in 1927, that mass production precedes mass consumption and alone makes it possible – first we figure out how to make millions of cans of soup. Then we have to figure out how to persuade people to buy canned soup instead of making their own.

Ask a group of people, as I have done many times, what was the main value behind their decision to purchase a can of soup, and you will be told "convenience." The successful marketing strategy behind canned soup has been to convince consumers it is more convenient. At the level of ingredients, if people do not grow their own, then there are issues of both cost and availability of materials. If people are not accustomed to cooking, then there is the additional barrier of learning how to make the soup, as well as the preparation time involved. Or you can simply buy a can of soup, open it, add water or milk, heat and serve.

Yet again, all of these conclusions are by choice, not by necessity. It is not difficult to get ingredients, even if you don't grow them yourself. Even if there is a short growing season, you can always preserve the tomatoes in some way toward future soup production. If you make more than one bowl at a time, then homemade soup may itself be frozen against future use, when it too needs only to be warmed and eaten, significantly dropping the cost per bowl in terms of time and money.

When it comes to making the soup, the skill level for making a basic soup is not much more than the ability to wash and chop vegetables, mixing them with a soup stock and cooking them up together.

What our can of soup reveals is the reciprocal relationship between technology and society, and between technology and culture. Society creates technology – which, in its turn, changes society. Culture chooses technology – which, in its turn, changes culture.

As more people left the farm for the city, there were fewer producers and more people doing other jobs than those involving agriculture. Add to this the advertising about the need for leisure – and whatever it is, it's not cooking! – and space is created for the canned soup that advertisers at first labelled "soup just like mother used to make." As the decades moved on, it became "soup just like grandma used to make." Eventually, when no one could remember great-grandma's soup, it just became the obvious choice for lunch.

Cooking skills, like many other types of practical knowledge, are essentially the product of oral culture. There are recipes, but how one follows them – the knack of rolling pastry or whatever – is passed along in the kitchen from generation to generation. If you have never made bread, a recipe is small help without lessons in what to do. It requires practice and experience – either acquiring your own or learning from the mistakes and successes of someone else.

In the kitchen, as in so many other parts of our lives, we are experiencing the hemorrhaging of practical knowledge, the technology acquired by trial and error over thousands of years across all ethnicities and cultures. You simply cannot learn to bake bread from the internet, nor how to swing a hammer for a day without hurting yourself. Our culture has increasingly replaced physical and experiential knowledge with the "I feel lucky" result of a computer search engine.

Combine the loss of cooking skills with the inability to can, pickle, salt or preserve that was commonplace in the 1950s and early 1960s, and we have created a generation of people for whom canned soup seems like the only option. They therefore *feel* they have no choice but to buy it – which makes the soup manufacturer very happy – but again, this is a choice, not a necessity.

Use advertising to push the ideas of personal choice and individual consumption – and watch communal cooking or even shared family meals become anachronisms. We don't need a pot of soup anymore, just a bowl of our favourite kind, and the can (which sells for the same price as before) now only has to be half as big.

We could make a similar case for the ways in which consumers have been trained to purchase fresh fruits and vegetables year-round instead of growing their own or preserving things when they are in season. Even in the smallest apartment, we could grow some produce in a window box all year round, making us less vulnerable to the inevitable disruptions in supply from a distance that the future will bring.

I've said a lot about soup here, but you will get

the point that this same kind of analysis can be applied to absolutely everything around us. When we reverse-engineer examples of our technology, we find the choices that led to them, the reasons for which the choices were made and the values that lie behind the reasons.

Sustainability is thus not a scientific or technological problem; it is a social and cultural problem.

When we unpack the values embedded in our technologies – and in our technological systems – we find that they are inextricably linked with the values expressed in our social and cultural systems.

To say we live in an unsustainable society, therefore, requires us to look at all of the systems together, not just to identify the bad technology or the poor choice in a particular situation. There is also no point in looking for culprits or villains, because we are all complicit in the society and culture in which we live. We can't escape it, any more than we can escape the air we breathe.

What we can do, however, is see the problem in terms of system design. In those linear, mechanical systems that underpin our society, there are inputs, processes and outputs. If we don't like the outputs, then we need to change either the inputs or the processes. Otherwise, the predictable and repetitive outcomes we expect from the Machine will continue and so will our unsustainable society.

And if our values are inputs into the technology we choose, then we have to either change our values or reorder them if we want different results.

Otherwise, nothing changes. How could it?

Chapter 5
A Problem of Design

A sustainable future requires better system design

Many people have described sustainability in terms of a three-legged stool, where one leg is the environment, the second is society and the third is the economy. The theory, of course, is that such an easily visualized model will enable people to better understand the interrelated elements necessary to effect changes toward a sustainable future.

Unfortunately, this model is utterly wrong. It is more harmful than helpful, best left on the scrap heap of good intentions.

For one thing, it is static. Obviously, stools aren't alive. A static model of a dynamic reality doesn't explain very much, especially if the whole problem is the failure of people to understand the dynamics of sustainability in the first place.

What is worse, the three equal legs to the stool skew the understanding of sustainability in the wrong direction. Making the economy equal to the environment and society leads to the absurdity that we would like to have a planet on which humans are able to live, but just can't afford one.

This reminds me of a scene at the end of the 1933 novel *When Worlds Collide*, by Edwin Balmer and Philip Wylie. A rocket is about to be launched that will try to transfer a few humans from our world (which is about to collide with a large rogue planet and thus be destroyed) to a companion planet that will be captured by the sun's gravitational field and remain as a new Earth. Just before liftoff, the richest man in the world shows up at the launch site with a planeload of American currency and (unsuccessfully) tries to buy his way onto the rocket.

One of biggest system design problems is how poorly we understand natural capital – what the Earth provides – and the cost to the planet that results from how we choose to live. It's not that we can't afford to spend the money we need on what is required. We simply choose to spend it on something else. In the same way, those people who are too busy to make soup, or whatever other task has been set aside because of a lack of time, choose to spend time doing other things they consider to be more important.

Check out the graphs of annual global military expenditures and compare these to the cost of what it would take to solve persistent problems (like clean drinking water or sanitation – pick your favourite unmet Millennium Development Goal!) – we have the money. Check out graphs of time spent playing video games, surfing the internet or being passively entertained by television programs or movies – we have the time too.

⁂

Economic analysis is only a tool. The things we analyze in economic terms and those we ignore are the result of choices we make that are the product of the values we hold, individually or collectively. Whether it is an evaluation of natural capital, ecosystem services, real cost accounting or other indicators beyond GDP (like happiness), we need to use our analytical tools in other ways, to different ends.

Economics is thus just a measure of our choices, a barometer of what we think is important. It is a way of measuring what it costs to choose one thing and not another. The question "How can we afford to do it?" needs to be balanced by a second question – "How can we afford not to do it?" – if we are to reach an appropriate conclusion.

Every culture back to the Tower of Babel has decided what it valued most when it came to large-scale projects. We could not afford to build the Pyramids any more than we could build the Great Wall of China; the cost of medieval cathedrals with all their intricate designs would drain our treasuries even if we could find the skill to make them.

Even now, we wonder how we could rebuild a railway system, but somehow it was managed in the nineteenth century, just as sewer, water, electricity and telephone systems were installed worldwide in the twentieth. We have difficulty filling the potholes, but fifty years ago a continent-wide highway system was built.

We've got the money, honey, and we've got the time too – we just don't use either of them any more wisely than the song suggests.

Sustainability without responsibility is impossible. In order for global society to be environmentally sustainable, it must reflect decision-making that is socially responsible on a global scale. This is the single biggest system design problem, because however "green" our local decision-making might appear to be, it does not seem to translate into a larger context.

We can't simply recycle ourselves into a better environmental future without making socially responsible decisions that benefit other people more than ourselves, the community more than the individual – decisions that result from a growing awareness of what we should do, and not just what we have the power to do at any particular moment.

As thinking humans, we need to understand the significance of our decisions about technology, not only for the society in which we live, but also for the environment.

The environment is the sum total of everything that affects our health and well-being. We tend to focus on how the environment affects us physically, but our environment also has psychological and – as many people would argue – spiritual effects on people.

Some of these effects are immediate and obvious – for example, standing out in a thunderstorm holding a metal pole likely means getting hit by lightning – but other effects are long term and more subtle. The steady increase in rates of previously rare kinds of cancer has been linked to human-made chemicals in the environment, but whether there is one culprit (like dioxins) or a combination of chemicals causing these increases is unclear. Similarly, increases in asthma, allergies and other immune system problems have been linked to changes in the environment over the last fifty years due to industrial and other pollution tied to the way we live in Western society.

We are what we eat, drink and breathe. We are no different than any other organism in terms of the way our environment affects our physical, psychological, emotional and – again, as many would say – our spiritual well-being.

Because of this, sustainability needs to be considered as a central element in our technology and in our technological systems.

While Western science and technology reshaped the planet in many ways, the longer-term effects are making the Earth less hospitable to existing life forms, including humans, than ever before.

We are not just changing the face of the

planet – in many places, we are making it uninhabitable by the very creatures that have lived here for thousands of years, including ourselves.

In Western culture, we seem to value our ability to change nature more than our ability to live with nature. The idea that nature is little more than clay, molded or shaped as we choose, passes almost without comment in Western culture, but it is not shared by other cultures (such as aboriginal cultures). When we think of the devastating consequences of our transformations of nature, we should be more hesitant about making irreversible changes – and more humble about our intentions.

One of the other values reflected in Western technology is the desire to improve things for the current generation, to make our lives easier, wealthier, longer or whatever. Placing such a high value on ourselves, however, means making choices that make life harder or poorer for other people, not just elsewhere in our global community, but in the future.

Using up non-renewable resources means they are gone – they will not be available either for us or for other people in the future. Given the fact that our environment is a planet with definite boundaries in space, we should realize that our resources are finite and limited, but most of the time that isn't reflected in the decisions we make as individuals and as a society.

The additional problem here is that the whole discussion tends to revolve around the human, putting people at the centre of the decisions about how we should live toward a sustainable future. This kind of anthropocentrism is necessary to a point, but if it becomes our only concern, then we will not have a planet on which anything familiar is able to live. A systems ethic requires all of the systems to be considered, not just the human ones – which is why we must understand we are only a very small part of the ecological system that sustains life of all kinds on Earth, in balance, as it has for more years than we can accurately count.

Every step we take leaves a footprint that affects the Earth as a whole. We can either step thoughtfully and knowingly, or blindly and foolishly.

Often this sort of discussion leads to a point where people are accused of being against technology, of wanting to go back to some simpler time when there was no technology.

Of course, we now realize this is the wrong response. The Luddites were right, but their response doesn't help us. The question is not whether to abandon technology, but what kinds of technology to choose – and what values are reflected in our choices.

Yet when it comes to understanding the global environmental system in all its complexity, no one can predict exactly what is going to happen or when.

From a common sense perspective, however, we know we shouldn't breathe the fumes out of a tailpipe or drink dirty water, just as we know it is not smart to step in front of a moving bus.

When it comes to using energy or consuming non-renewable resources, the same common sense perspective should apply. Living sustainably means making common sense decisions about how you live, not simply picking the right expert with whom to agree.

If your food comes from far away, what happens if something prevents the trucks from arriving? Why not grow some of your own? If you need clean water to drink, what happens if everyone pollutes it the way you do? If clean air is needed for our health, then why would you hesitate to fix the engine on your car so it stops burning oil – or perhaps put away the car and ride a bike?

These are the kinds of common sense decisions that have made it possible for humans to survive (and to thrive) before. Our society is environmentally unsustainable because we have stopped making these kinds of decisions, decisions that reflect responsibility toward the planet of which we are a part, responsibility toward other living creatures and responsibility toward other people, both present and future.

It is not our technology by itself that constitutes the real problem confronting global society today. It is the interaction of our technological systems with social and cultural systems, the way in which certain types of technology are co-opted or preferred – usually as a means of acquiring power – that has created the dilemmas confronting citizens of this new century.

The archetypal person in the loincloth, the representative of a supposedly "primitive" culture of earlier times, looks at our inability to solve the simplest and most basic of problems (despite all our shiny new tools) and shakes his head in disbelief at our ineptitude.

Technology itself is not bad, but it is also not accidental. What we do, we choose to do; what we make, we choose to make. Our problems, therefore, lie not with the technology we have at our disposal, but with the social and cultural structures within which we make these choices and that guide the technology we pick.

There is no such thing as autonomous technology; while systems may be set up to operate with a minimum of human intervention, they still are not independent. To claim technology is a runaway train over which we have no control is merely to abdicate responsibility for the choices that we all make, and continue to make, as the train picks up speed – and this assumes it is a good idea to think of it as a train at all!

Sustainability – both in terms of society and in terms of the environment – needs to be the central value in the choices we make, as individuals and as a culture, one that is therefore reflected in all of our systems.

It's not enough to make money. We have to live to spend it. It's not enough to live well today. We need to live well tomorrow. In a global community, it's not enough to make decisions only for our own benefit. We need to make decisions that benefit other people – so that they will be encouraged to make decisions that, in their turn, will benefit us.

We need to bring out into the open the choices we are currently making about the tools we use. We need to be aware of our choices and our responsibilities, not in order to turn our backs upon technology, but to choose what kinds of technology we need in order to create a sustainable future for ourselves and for future generations.

Every day, we make decisions as individuals, as members of Western society and as participants in 21st-century consumer culture. These decisions reflect our values. We need to consider those values carefully, keeping what will create a sustainable future and discarding what threatens to destroy it.

❧

At the risk of returning to the mundane with a crash, this brings me back to soup. It is one thing to exhort and persuade. It is quite another to land the system design problem right back in the dirt from whence it came.

If you were uncomfortable with the soup can analysis or were unhappy with the values I identified behind the choices that culminated in the bowlful you just ate for lunch, consider this:

You could stop eating soup and eat something else instead. Or you could learn how to make your own soup from scratch – not hard, and something that (like any other cooking skill) becomes easier and more sophisticated with practice.

You would have to find fresh ingredients in season, or use frozen ones out of season. You would discover the difference in taste between produce from your own garden and things from a distance – and would quickly realize that a couple of zucchini plants is enough! You would find out how easy it is to grow herbs – and learn that dill weed really lives up to its name.

Your kids, meanwhile, are watching in amazement, as they learn that soup does not have to come in a can from the store, that the yard grows more than weeds and grass and that chickens are not raised in cellophane.

All your soup now needs to be eaten, so it is either packaged for the freezer (and displayed to your colleagues at work or at school who are still ruefully eating canned stuff that neither smells nor tastes as good) or leads to invitations to friends to join you for a shared meal.

As the soup movement grows, more gardens are planted, more people understand their relation to the food they eat and how it is produced, more people learn how to cook things for themselves, fewer people eat alone, money is spent close to home instead on food imports and everyone is healthier as a result.

Eventually, the soup manufacturer realizes that sales are dropping and – after the failure of an advertising campaign focused on the hygienic dangers of home cooking – shifts to making some other product to satisfy the shareholders, many of whom are no longer interested in canned soup anyway.

One person, one choice at a time, can change the system. Complex, interrelated systems can compound problems into nightmares, but they can also be used to multiply personal choices into realistic and practical answers toward a sustainable future for us all.

Chapter 6
System Ethics

System choices require intentional system ethics

"This bottle is our property. Anyone using, destroying or retaining [it] will be prosecuted. Blackwood Bros."

"This bottle is our property. Any charge made therefor simply covers its use while containing goods bottled by us and must be returned when empty. Blackwoods Limited."

Twenty-five years ago, my parents had to dig up the old septic tank and install a new one. In the process of the excavation, an even older trash pit was uncovered, yielding all sorts of glass bottles that had survived for the better part of a century underground (no doubt the soup cans had rusted away!).

I kept two of the bottles, soda bottles from

Blackwoods Beverages from the 1920s or earlier. For me, having grown up with bottle drives (where we collected pop bottles to cash in the deposits to fund sports teams and Boy Scouts) and then lived into the era in which glass became plastic (and was not returned but recycled instead), these bottles are a novelty.

The fact that the company asserted property rights over the container shows how social and cultural values can shift over time. Misuse or fail to return one of their bottles and (in theory, at least) you could be arrested!

Packaging that is more valuable than its contents is uncommon and certainly shifts the production/consumption system away from both excess packaging and unnecessary waste.

It also reflects how values very different from the ones we hold today can be embedded in similar social and cultural systems. Things change, but not always for the better.

To change our unsustainable global society we need to solve the problem of poor system design. That means addressing the decisions we make and identifying both the reasons that lie behind our choices and the values that lie behind the reasons.

A new global system of ethics – or a systems ethic – that deals with the whole problem at once is the only workable response to making choices about a system unsustainable in its design and operation. This ethic must incorporate principles of sustainability that people from all places, cultures and ethnicities can understand and accept. It needs to radically change the way they make daily decisions about technology, and it needs to foster a deliberate and intentional attitude toward everyday choices.

Such an outlandish and impossible statement requires some explanation.

If ethics is the nexus between technology and sustainability, we can neither achieve sustainability nor understand technology apart from the realization that ethics is crucial, not accidental.

First, we are all moral agents – we have the means, the opportunity and the responsibility to make moral choices. What we consider to be good or bad reflects our personal morality; how we apply our morality to the decisions we make reflects our personal ethics.

Second, our personal agency ends when our choices conflict with the choices of others. Our freedom is bounded by the freedom of other people, each with a right to enjoy his or her life equal to our own. Society then consists of a social consensus (I hesitate to use the word *contract* outside of a legal context) as to the bounds of acceptable behavior that limits our freedom to act.

We also place restrictions on our behaviour with respect to moral patients – those who have moral status but are acted upon rather than moral actors themselves. (We could include children here, or adults with cognitive disabilities that limit their ability to choose for themselves. Some philosophers even extend moral patiency to non-human organisms, arguing this is the basis for restrictions against cruelty to animals.)

One of the key problems with our unsustainable culture is that our choices today conflict with the choices of future generations. It is not enough to understand or even to implement a system of ecological justice that balances the needs and desires of the current generation around the world, in all its diversity and circumstance. Our systems ethic has to balance our choices against the choices of future generations – and not just the human ones either.

Third, none of the existing perspectives from which people view ethical dilemmas are helpful in solving the system design problems we have identified.

A consequentialist approach is completely undermined by complexity and our inability to identify and understand all the system components. Because we can't predict the future, it is hard to choose between outcomes when we are only choosing between probabilities that may or may not lead to the outcome we expect.

A principle-based approach that does not consider consequences, however, lacks the flexibility needed in widely varying situations. One size never fits all, and principles applied without considering consequences risk being catastrophic if the axioms behind them are incorrect or unsustainable.

While a focus on the ethics of personal virtue might seem a more promising alternative, it would require a global consensus as to what those virtues might be; what it means to be a "good" person means different things in different cultures.

A system design problem that results from poor

ethical choices, both by individuals and by society, requires a systems ethic. You simply can't isolate specific problems within a complex system and hope to apply an individual analysis and response to each one. Problems will always appear faster than they can be solved, for each partial solution risks being the cause of yet another set of other problems.

⁓

We need to choose wisely, because our technology extends our reach further than ever before, in terms of both distance and time. What we choose to do not only affects people on the other side of the world, but it can also affect people who are not yet even born.

Consumer culture does not encourage this kind of extended awareness. Consumerism is self-focused, driven by individual needs that are infinitely elastic (which means there is no necessary end to what you might want or "need"), and calculated in financial terms that usually do not factor in environmental or social costs.

Our needs may seem infinite, but our resources are not. The realization of the limits to our resources pushes us to ask "should" questions in making ethical decisions, not just "could" questions: "I *could* buy that, but *should* I buy it?" "I *could* do this to someone else, but *should* I do it?" Asking the "should" question takes the decision out of the realm of power and puts it where it belongs – in the realm of responsibility.

Making socially responsible decisions with scarce resources (our own resources or the planet's) encourages other people to do the same. After all, when we are powerless, we would want those people with power over us – whether they are next door or around the world – to think about whether they *should* do what they have the power to do, or not do, as they choose.

Social responsibility, therefore, requires us to understand the ethical choices presented to us every day and to make wise decisions in both our private roles and our public lives. Without socially responsible decisions, made by all of us, ecological sustainability becomes an impossible goal.

🐜

A systems ethic requires consideration of all the stakeholders, beginning not with options for action nor reasons to choose one course or another, but with the values inherent in the situation. We need to start with the values, what we think is important, and only then move to the next step of identifying the reasons we have to make the choices we must if we are to travel in the direction we want to go.

What is important to all of the stakeholders, present and future, human and non-human, organic and inorganic, that share the planet and all of its systems with us?

A systems ethic involves personal ethics, but recognizes the community context within which such personal choices take place. It gives voice to all the people (both present and distant) who are affected by whatever decision is made. It may start with anthropocentric concerns, but the circle is widened to include all the other characters and contexts.

Personal ethics are a necessary start, because they entail ownership, responsibility and engagement, but that is insufficient. We are also in relationships, whether as members of a family or a community. This means that our personal ethical choices are framed – though not determined – by those around us. This community context is not only present and local; it is also distant in space and time, including people on the other side of the world as well as future generations.

A systems ethic thus also involves social ethics, but in a way that the larger social values are always contextualized within local situations. Whatever larger principles or issues are at stake, the decisions affect specific individuals in particular locations and are always understood, finally, in local and personal terms.

A systems ethic never overlooks the fact that these larger principles are lived out locally in the lives of those affected by the decisions that are made. ("Think globally, act locally" is one reflection of a systems ethic.)

A systems ethic, therefore, is also reciprocal, realizing that the decisions and their influences on future relations never move in only one direction. Reflexivity, reciprocity, give-and-take – whatever term you use to describe it, in dynamic systems, complexity is never linear.

⁂

Living responsibly requires an understanding of how society is shaped by technology, and how society makes choices about the type of technology it develops and uses. It also requires an awareness of the values reflected in the choices we make as individuals and as a society, and the ability to discern the good and bad consequences of those choices.

These two concerns need to be expressed in terms of both ecological sustainability and social responsibility. The decisions we make must be ecologically sustainable in the longer term, or we will change the planet into a place that will no longer support human life in the locations (and numbers) we have at the moment – a future in which there are even more people requiring resources that have become even more scarce. More than ever before, the choices we make at this time in the history of the planet will determine what kind of a future there will be for all the people on Earth.

These ecologically sustainable decisions also must be made in the context of social responsibility, because a world divided into the rich and the poor can too easily be divided into those who will live and those who will die. Justice and fairness therefore must also be key elements of globally sustainable decisions, or those people whose interests (and perhaps even survival) are ignored are not likely to support whatever is decided. As people living in one of the richest countries in the world, we can't expect the burdens of reversing climate change, for example, to be carried by those for whom every day is a struggle to survive because of a lack of food, water and the basic necessities of life.

So where does this leave us, in making choices about technology for ourselves, for our society and for future generations?

We can't escape the need to make these choices, but we should be wary of easy answers. When you are dealing with the complex relationship between technological systems and social and cultural systems, a simple and straightforward answer is unlikely. Factor in the reality that our choices are not just for ourselves as humans but are for all the children of the Earth – and that however complex we realize our own systems might be, ecological systems are much more sophisticated and complex – and any answer offered in mechanical terms is merely inane.

Once is not enough, nor does one size fit all. Because we are dealing with dynamic systems, not static ones, our decisions will need to be made differently, over and over again, as the circumstances of the interrelated systems change.

❦

Yet for me to say we need a new systems ethic means that I have something of an idea of what we need to do.

One facet of this new systems ethic is found in *Gift Ecology*, in which the values associated with an exchange are replaced by the values associated with a gift, acknowledging that we live in a universe of relations and not an environment of connections. Economics needs to return to *oikonomia*, caring for what is passed to the next generation and subordinated to the ecology within which we live as creatures of Earth.

But it is not enough merely to recognize that we live in a universe of relations, laden with values. We also need to live and to act. Moral choices lie at the heart of many of our decisions. These moral choices are heavily influenced by our society, our culture, our religious beliefs, our personal family history and our own experiences.

A new systems ethic must be intentional in the moment, aware of the context of eternity but alive and in relation as our decisions today link the past with possibility. Our ethics must no longer be accidental, nor should our choices be made unawares or without thought – or by deferring responsibility to some larger organization (like the government) without direct and immediate accountability to us for what is decided.

Our values lead to reasons for choosing one course of action over another. Understanding those values, knowing and articulating those reasons, is the only course to an intentional systems ethic that brings all of our abilities, experience and wisdom into focus for the challenges that confront our generation.

The tools we use and the tools we need to use are the ones that are appropriate to the situation and to doing what we have chosen to do. Technology is not outside of us, something separate and distant, a force on its own. Technology is always within our grasp, because it reflects an understanding of how the universe works that we have found to be useful and practical. The objects of technology that we wield, within the technological systems we have designed and built, are expressions of what we know and understand – used toward an end (or for a purpose) that we want.

To return to the beginning of the book and to

restate this point more simply, technology is in our heads, not merely in our hands.

❧

We can be certain that other people will not have the same perspective as we do on moral choices. In a diverse society like our own, there must be a framework of respect and discussion to enable our own voice to be heard and for us to respect the voices of other people who have different opinions.

Most of us, whether we like it or not, live and work with other people. Ethical decision-making therefore does not happen in a vacuum. It happens in community. Our choices affect more than ourselves and so they need to be made with other people in mind.

They also need to reflect the ecological reality that we are part of the planet itself, not separate from it.

In other words, we also live in a complex web of interrelated systems – whether they are ecological, biological, social, cultural or technological – whose interrelationships are changing all the time. We try to manage it, one disconnected piece at a time, pretending to have some knowledge, understanding or control over some limited aspect of a global system far too complex for us to comprehend.

Yet we can't manage organic complexity simply by imposing frameworks geared to crude models of linear causality. There are far too many elements, far too many variables and inadequate predictive tools to make this work.

What's worse, we approach this design problem with a combination of arrogance and presumption, not with humility and a willingness to listen either to each other or to what the planet has to say – but that's what we need if we are going to incorporate the same kind of resilience that we find in natural systems into the systems we design ourselves.

We also need to realize that emergence is a natural design feature that we have to weave into systems of our own creation. Organic systems demonstrate this principle, essentially defeating entropy by increasing both organization and complexity – and then demonstrating emergent characteristics or new properties over time. This means we need to design *with* nature rather than against it – an approach that would reflect a major value shift in a culture that, for hundreds of years in the West, has determinedly imposed its will on the natural world to suit its own purposes and whims.

We mistake force for power and think ourselves superior to aspects of the natural world whose power is more subtle and is reflected in ecological rather than technological systems. Whether it is the loss of honeybees to pollinate crops, the disappearance of the ants and earthworms that work the soil, or the death of coral reefs that fix carbon and regenerate the oceans, we are slowly learning about the complexities of ecological systems, now that it may be too late to alter the consequences of the poor choices we have made.

The new systems ethic therefore must be based upon principles, not probabilities – we cannot predict the outcomes of system changes if we do not adequately understand all the variables within the system.

Within that universe of relations – all of our relations, not just the human ones – we need to demonstrate by our choices what Albert Schweitzer called "a reverence for life." We need to live with respect in Creation – realizing that whatever its origins or however the world around and within us has been created, we are neither its designer nor its Creator.

Whatever its origins, however, the future of that Creation depends on the choices we make. All around us, every day, we are literally choosing the future. We are crafting the story that our children and all the children of Earth will experience for themselves.

We are choosing that future right here, right now, within the embrace of our arms and under the soles of our feet.

We need to think about what we choose and why, allowing more important values to guide our choices than the ones we have tended to use before. Those individual choices have created the world in which we now live, with all its problems and potential. If we make better choices today than we did yesterday – not brilliant choices, just better ones – the world shifts, however slightly, toward a sustainable future for us all.

A new systems ethic – grounded in a respect for Creation, a concern for other people and a reverence for life itself – is within our reach, if we choose it to underpin all the decisions that we make.

In everything we do, every day, there is both responsibility and possibility – grounded not in despair, nor in denial, but in hope.

Postlude
The End of Progress

I have often wondered what it will take to change the way people see each other and the world in which they live. Perhaps a recollection is needed instead, something out of ancestral memory that recalls how humans used to live in relation with the Earth.

Some say it will take global catastrophe, something so extreme that no one, anywhere, can dispute the necessity of changing the way we do things. This makes sustainability into a global game of *Survivor*, in which those who are left will try to put together the pieces. Otherwise, it seems people assume that bad things will happen to someone else, not to them, and that they will be among the few who escape catastrophe.

It's like waiting for the ship to hit the iceberg so that the few who make it to the lifeboats can row toward the possibility of an island where they might be able to live happily ever after. It also presumes that we will be among the lucky ones who make it to the lifeboats.

Others say that society never will change, that all we can do is grittily hang on to what we can as the world around us heats up into a nightmare without end. We must prepare for the worst, because civilization is a veneer of convenience easily stripped away from the animals we really are.

Yet I refuse to believe that thousands of years of social development, the fruits of the spirit that have taken humans beings through the absolute worst of circumstances, will be so easily discarded that savagery is our only option even in such a future.

For myself, I think change will come as quickly as a rabbit becomes a duck. If you have ever seen those line drawings that illustrate a gestalt shift, you will know what I mean. The picture is a rabbit, can only be a rabbit, until the moment when your brain rearranges the lines and the rabbit forever turns into a duck.

We live in an unsustainable society, locally and globally. Activists protest, prophets foretell disaster, analysts recount the dismal numbers and still nothing changes. Governments continue with business as usual, and business as usual continues.

But I think there is a sea change about to sweep the planet – and not because I am engaged in some wish-fulfillment exercise.

I watch parents work extra jobs and reduce the number of children they have because they can't afford to have any more. They are sacrificing their retirements for college and university education; for the gadgets and trappings of the consumer lifestyle; for sports activities and music lessons; and a host of other things for their children.

But as they spend $30,000 a year on hockey equipment, ice fees and travel for little Johnny on the off chance he might make it to the National Hockey League, I wonder what will happen when they realize that their lifestyle means Johnny will grow up in a world without ice. What will those parents think of their own sacrifices when they realize that, if Johnny ever has any kids of his own, he will have to watch them suffer from hunger, or thirst, or heat, or any of the myriad environmental disasters that most places on Earth will experience as the climate shift continues over the next 30 years?

What happens when those 30 to 50 per cent of the global population currently under the age of 30 realize that they are being educated for a future that will not exist, to do jobs that will no longer be available, that the promise of a future better than the present is hollow because of what their elders enjoy at the moment?

What happens to the economy when people everywhere realize they are investing their money in a global Ponzi scheme, in which the promise of future returns is a sham because it is based on fictions of the marketplace and not on the reality of planetary boundaries?

What happens, in other words, when the rabbit of consumption forever becomes the duck of unsustainability?

It reminds me of the sea change that happened around smoking.

As a teenager, I grew up in a culture in which smoking was the norm. I was a non-smoker and so was always treated with condescension as I declined cigarettes and endured rooms – including classrooms – full of second-hand smoke. The majority of men smoked. Smoking was inevitably part of social events, from dances to bars to restaurants and coffee shops.

I remember at university, in a seminar room without ventilation, students would turn the No Smoking signs yellow with cigarette smoke and the professor accepted this as the norm. After one particularly noxious class, I protested bitterly to a friend in the Canadian militia, saying I needed a gas mask to continue in the course. He unexpectedly offered to loan me his military-issue double-canister gas mask, which offer – to save face – I had to accept.

The next day, it was a smoke-free class almost to the very end, when people decided to light up after all. The professor turned to the blackboard to write something, and so, behind his back, I put on the gas mask.

When he turned around, it was the only time I ever saw him at a loss for words. My classmates, following his stunned eyes to see me sitting there in masked judgment on their behaviour, were furious, stubbing out their cigarettes and cursing under their breath. Recovering somewhat, the professor said, "You have made your point, Mr. Denton. You may take off the mask now."

Ever the pacifist, I replied: "I don't think so. There is still smoke in the air," and left the mask on for the remainder of the class.

When I returned to teach in the same department, 25 years later, I discovered that I was still referred to by the nickname that I had earned that day: "Smokey."

Smokers are now in the minority, though it seems women are as likely to smoke as men. Admitting you are a smoker today means admitting a kind of moral failure, accepting your banishment to a circle of shame a prescribed number of metres away from any building entrance to indulge your addiction, regardless of weather. There is no smoking in bars or at social events; coffee shops and restaurants no longer have a only a couple of hypothetical No Smoking tables; hotels punish those who smoke in their rooms; and every few months there are murmurs of laws to forbid smoking in vehicles when children are present.

The rabbit has forever become a duck, not because of anti-smoking education, warnings on cigarette packages, increases in taxes, more medical evidence related to cancer or any amount of public harangue. All these things had been tried for decades with nothing more than marginal success. At some point, a critical mass of people decided – or realized – that smoking was socially unacceptable. The moral narrative shifted as a result.

I see the same thing happening with sustainability.

The lifestyle of excess, of hyper-consumption and waste, is under threat – which is why so much money continues to be spent on advertising. Consumer debt rises as we buy new things we didn't realize we needed, with money we hope to have someday, but many people are deeply anxious about the future.

We have been lectured for decades to "reduce, reuse and recycle," subjected to environmental levies, warned about the health effects of toxic chemicals in what we drink, what we eat and where we live. At a macro level, little has changed; but it has been interesting to watch what happens when social pressures start to build.

I remember years ago witnessing the rollout of the blue box program in Toronto. Toronto city council made available several thousand free boxes for curbside recycling that were snapped up in a matter of hours. The program snowballed, as the boxes appeared along every sidewalk over the next few months. Those residents without boxes were harried until they too had one by the curb – and I watched neighbours checking each other's performance to see what had been recycled this week.

At the same time, in the space of three years, the rural community where I lived went from collecting newspapers (hazardously stored in one side of the garage at the church manse) and selling them for $50 a ton, to pleading with someone to take them away for free, to having to pay for their removal in the third year. Newspaper recycling instantly outstripped any local capacity to actually process the paper – and it was the same for cardboard, glass and plastic.

The rabbit forever became a duck, even though the recycling piled up with no place to put it.

So when I am challenged about timelines, about how slow and fruitless the process of change toward sustainable lifestyles seems to be, about whether we have enough time to do what needs to be done, my response is a simple question: "How long does it take for a rabbit to become a duck?"

The answer could be "It never will." But it also could be "In the blink of an eye."

Reflecting on the nature of technology and our choices inevitably leads us to think about the word *progress*. We find the same dilemma here with the use of capital letters: do we mean to reify it and turn it into some grand, concrete idea of "Progress," or are we just using it descriptively, meaning "progress"?

Go back to the discussion about the Luddites and how their story has been co-opted into our culture, and you would have to say it was a capital *P*. The Luddites were not just opposed to changes in their lives; they were represented as opposing Progress. People who raise concerns about new technologies are often labelled in the same way – even caution or doubt is on par with rejection.

Progress became the ideology driving the machinery of industrial society in the 19th and 20th centuries. Doubts were not easily entertained – and where they existed, they certainly were not doubts about the system as a whole. Whether it was in the Soviet Union, China or the United States – with their proxies and allies in between – public doubts that the system was moving inexorably to a better future for all were handled harshly by the authorities in all three countries.

You might have been asked, "Are things better today than yesterday? Will they be better tomorrow than today?" Answer either question in the negative, or raise doubt about the validity of the official response, and everything – including your life – was potentially in danger. One might argue that punishing dissent was not about politics – the danger to the regime was when people questioned the ideology of Progress.

This is where the story has to change. Getting away from the capital letters, progress is a measurement. To measure anything you need a benchmark – a place to start – and a unit of measure. But attempt to measure "progress" in any technological system, however, and you run into trouble when social, cultural and ecological systems are also considered, as they should be.

There is difference and change, to be sure, but every technological system has its advantages and disadvantages. It must be appropriate to the society in which it is developed and used, or that society can get into trouble very quickly if its technological effects overwhelm the natural resilience of the other systems.

Progress (with the capital letter) assumes the

forward, linear march associated with the perpetual growth model of contemporary economic forecasting. Yet outputs only increase as inputs increase or processes become more efficient. This is completely predictable in a mechanical production system, but it is incomprehensible in an organic system that is inherently circular, in which things are born, grow, die and decompose.

A sustainable future, therefore, is simply impossible if we don't change the line to a circle, the machine to an organism, the production and consumption system to one of growth and regeneration.

The end of Progress is the key – or perhaps this time it should be the End of Progress. What is its purpose, its end, its reason for being? As an idea driving the changes of the industrial culture from its roots in the 18th century through the 20th, Progress saw frontiers of all kinds – physical and intellectual – pushed back in its name. But if the end of Progress was to build a better future for all people, then it is no longer a useful idea in our current, unsustainable situation.

That end or purpose now may be served only by making sustainable choices – not by driving forward toward a new frontier, but by recollecting and remembering, circling back to recall our relations with each other and with the Earth. It means rediscovering and reimagining whatever enabled other cultures to survive and thrive for thousands of years before us.

✧

If I could pick one technological object as both a symbol of our unsustainable culture and as a barometer of change toward a sustainable future, it would be the disposable plastic water bottle.

The values it embeds are wrong at so many levels. Its persistence is almost entirely due to the choices of individuals who fail to recognize these values for what they are and so fail to make better and more sustainable choices.

As the disposable plastic water bottle goes, so will go our civilization.

Such a pronouncement requires some justification. A brief systems analysis, reverse-engineering the values embedded within it like we did with the soup can, easily demonstrates what is at stake.

Bottling and selling water means that water is no longer a basic human right. The commodification of water – the assertion of private property and ownership rights – means that through such commercialization water is no longer something that is part of "the commons." Given that water is essential to life, commodifying it in effect creates a life tax – if you do not have the money to buy it, you die.

It is also entirely anthropocentric. If we commodify water, we remove it from the natural cycle in which other systems besides human ones require it, prioritizing our needs over those of other creatures. The ironic outcome of these choices, moreover, is a declining hydrological cycle that makes even less water available the following season; when we remove water from that cycle by bottling and shipping it elsewhere, it obviously does not return.

Drinking bottled water brings with it illusions of safety that fly in the face of easily available evidence. I recall a university microbiology course that for many years began with a lab in which water bottles brought from home were tested against commercial ones, and then various water sources around the campus were tested too. Consistently the highest bacteriological contamination was found in the commercial water bottles, while the lowest – and therefore the cleanest water – was found in the toilets!

The high financial cost of drinking bottled water has social implications too, for the same money spent by individuals on bottled water in places where the municipal supply is inadequate could quite easily be spent instead on making clean water available for everyone, not just those with the price of a bottle in their pockets.

Then there is the personal, social and ecological cost of the disposable plastic bottle itself. Water is a universal solvent, so the non-renewable petrochemical resources spent on making the bottle in turn potentially leach toxins into the water in the bottle at levels of concentration that municipal water supplies would not be allowed to contain.

Once used, the disposable bottle ends up in landfills, rarely recycled into anything useful or even into more of the same containers, because we have too many different types of plastics.

Washed out to sea, the rings and bottle caps become hazards to the seabirds that mistake them for food, while the dissolving plastic turns the sea into dead zones like those found in the Pacific gyre or turns into micropellets of toxins that are consumed by small fish, eventually bioaccumulating in the larger fish that we then eat.

Costing more per litre than gasoline, bottled water is often merely local tap water put into a bottle, not itself subject to the same stringent testing as municipal water supplies.

Yet despite all these obvious concerns, in the string of values embedded in this object, one value comes first: convenience. We have become convinced, even if shown the hazards of disposable plastic water bottles, that their convenience trumps everything else and that reservations about the water's purity are misplaced.

Making better choices today than yesterday might mean gradually reducing the number of disposable water bottles you purchase by at first refilling some of them – and once you are tired of bottles with leaky caps, you buy one that is supposed to be refilled. In a week of such changing choices, you could reduce your consumption from 20 bottles to zero, and the cost from $25 to whatever your refillable bottle might cost. Water is not addictive – there is literally no barrier to changing these choices, apart from habit, and such a shift results from simply making a better choice today than you did yesterday.

It is a small personal choice, but considering how many millions of water bottles are discarded in a year and the devastation they cause, such small choices multiplied together are what will either save or destroy the Earth.

🐜

One of the amusements of being a poor typist is that the words my fingers produce on the keyboard are fixed without hesitation by the computer (now with autocorrect as well as spellcheck). Many times the word *destroy* comes out as "destory." It is an interesting mistake, because it is precisely when we destory the Earth that we are more likely to destroy it as well.

The stories told around our cultural fires were stories of the Earth as well as of people: tales of the gods interacting with the elements; explanations for why things around us work one way and not another.

When we lose the stories that remind us of who we are and where we come from, we lose our way; when we lose our stories of the Earth, we lose what it means to have a home. The famous Apollo 8 pictures of the Earth from space reminded us, as a species, that the Earth is our home.

142

We need to weave new stories as well as recall old ones, for ourselves, for our children, and for the future that awaits those who will live into it. But the best stories are not those with clear endings, with neat conclusions, with a meaning easily expounded and learned. The best stories are the ones that end with questions, "to be continued," allowing the audience to write themselves into the plot, to become characters and actors, active participants in a story yet to be concluded because each scene creates more possibilities than the threads it ties off.

Our story, the Earth story, is far from over.

Postscript

"Better to pass boldly into that other world, in the full glory of some passion, than fade and wither dismally with age."

—James Joyce, "The Dead," in *Dubliners*

I always seem to think better in an Irish pub. So when I found myself in Washington, D.C., during Hurricane Sandy, with everything on the Hill shut down, the fact that the only place close by to eat was an Irish pub named The Dubliner seemed propitious.

The name, of course, reminded me of James Joyce's famous collection of short stories, which I had not read for some time. Drawn back to his notion of "epiphany," I found this epigraph and placed it in the eye of the storm in which I have found myself in the past year, of which Sandy was merely the latest and most physical example.

The impulse in January 2012 to fling out into the world what had long been brewing in my mind had led, within a few months, to the publication of *Gift Ecology: Reimagining a Sustainable World*. Two weeks to the day after the launch in October, I found myself in the middle of Hurricane Sandy, having come to Washington for civil society consultations with the United Nations Environment Program (RONA) on the follow-up to the 2012 conference in Rio de Janeiro (Rio+20). En route, I had participated in PowerShift 2012 in Ottawa, with hundreds of younger people who eyed my whitening hair with tolerant amusement.

With all this swirling about me, and with consultations cancelled and nothing to do, I found myself in the pub, pulling out the notebook to try and outline the book on technology and sustainability I had promised my students (and myself) for a decade that I would write.

So from afternoon into the evening, as the winds rose and the rain fell harder, that feeling of an impending epiphany grew. Pages were filled with scrawl and, just as I faltered toward the end, the first chords of the live music rang out, and the moment became what Joyce himself had described.

Things became even more clearly focused later that night when, at my request, musician Brian Gaffney recalled and performed a powerful song I had only heard performed live once before – in 1980, in a small pub on the outskirts of Dublin. The lyrics of "Only Our Rivers Run Free" drove home the point of Gabriel's epiphany:

We mourn what was, we wish for what we will never see, and so we miss the possibility of the moment.

To be human means to live toward possibility, to embody potential for growth and difference and change. In an atomistic age, it is easy to feel isolated and separate; in a mechanistic age, it is just as easy to fall into patterns of thinking and living that are linear and therefore give the impression of both control and predictability. In such a view of life, hope requires validation, as possibility mutates into probability and choices are weighed in the balance of anticipated (and immediate) results.

It is interesting to see social networking and the internet opening our local lives to possibility from a distance, from the outside. There is even more possibility through opening our lives to what we find within ourselves, as we travel inward to the source of Life and relation that finds itself expressed in everything from religions and spirituality to art, poetry, music and love.

All these things become creative expressions of the possibilities that humans embody, outside of the predictability of linear systems. Life is an unfolding dynamic, and sustainability is anything but a three-legged stool.

Embracing the epiphany of Gabriel, the main character of "The Dead," means living the day in light of what it brings, instead of lurking in the shadows of what we wanted it to bring. It means understanding what we say, think and do in the larger context of a hope found not in our own lives but within the universe of relations, a universe in which we are not alone, no matter how alone we might feel. It means living against the tide of an age in which denial is paired with despair.

To quote another Irish author, in William Butler Yeats's "The Second Coming," our age is marked by the fact that "The best lack all conviction, while the worst / Are full of passionate intensity." Ultimately, sustainability is about creating hope, not simply deciding what to do.

Measurements are retrospective; they are how

we count what has already happened. As a result, they are never predictive. The more variables in any situation and the more complex the system, the less we can extrapolate toward any future situation. Think about weather forecasting – as long as you can explain today why you were wrong yesterday, you keep your job, but even on the flat prairies, weather forecasts 12 hours out are often completely inaccurate.

So, to William and to Kaley, who worked through a hurricane because it was their job to serve customers like me; to Brian, whose music sealed the moment; and to those others bustling around the pub, whose names I did not learn – thank you for reminding me of the way individual people and their choices are woven within all of the systems of technology that shape our society and culture, whether we see them or know them or not.

This book, like *Gift Ecology*, is flung out into the world for you to read and to think about, not knowing where it will land or with whom it will find a home. I hope it will help change how people make choices about technology, so that all of Earth's children may experience a sustainable future – one that is ruled by passion for others instead of by profit for oneself; one that is governed by love and respect instead of by punishment and reward.

What you choose makes a difference, regardless of how old you are or whatever place you have right now in the world.

And you do choose, all day long, every day ... just as I do.

The question is not whether we choose, but what – and why.

— Peter Denton
Winnipeg, Canada
October 2014

Bookshelf

The background to this book reflects the same authors and ideas as are found "on the bookshelf" in *Gift Ecology: Reimagining a Sustainable World* (Rocky Mountain Books, 2012).

Adding to the items found there, I would emphasize the continued importance of Ursula Franklin's *The Real World of Technology* (Anansi, 1989; 2nd ed. 1999). While I fundamentally object to Kevin Kelly's *What Technology Wants* (Viking, 2010), it is worth reading simply to highlight how far we can go in the direction of reifying Technology into something beyond our control. W. Brian Arthur's *The Nature of Technology: What It Is and How It Evolves* (The Free Press, 2009) is more balanced, but does not advance our understanding of the subject as much as I had hoped, given the book's title. (I reviewed his book in *Essays in Philosophy* 13: 2 [2012], 583–89.)

In the larger context of thinking about the values that underpin our social and cultural choices, I have appreciated Roberto de Vogli's *Progress or Collapse: The Crises of Market Greed* (Routledge, 2013), along with Bill McKibben's *Oil and Honey: The Education of an Unlikely Activist* (Macmillan, 2013). Vandana Shiva's *Making Peace with the Earth* (Pluto Press, 2013) appeared after *Gift Ecology*, so I was unable to refer to it before now, but her work demonstrates how choices between social and cultural values around the basics of water, land and food are the real battleground for sustainability. Given the scale of the problems she describes and the size of the population involved, the struggle for a sustainable future might be won or lost in India alone.

Charles Eisenstein writes on topics for which I have much sympathy, though I wish he was more pragmatic about how the value shifts he sees as crucial for a better future might be brought about. Michael and Joyce Huesemann's *Techno-Fix: Why Technology Won't Save Us or the Environment* (New Society, 2011) makes explicit the problems that I think stem from a misunderstanding of the nature of technology, though their solutions (like their analyses) tend to falter because they involve a reification of terms like "Technology" and "Democracy."

Over the years, I have lost count of the conversations with colleagues, students and friends on the subjects in this book that have shaped, challenged and added to its argument. If technology is in our heads, then that kind of intellectual commons is where we all glean what we need for the choices we have to make. The community is stronger for the stories we share, around whatever cultural fires our generation tends toward a sustainable future.

RMB saved the following resources by printing the pages of this book on chlorine-free paper made with 100% post-consumer waste:

Trees · 6, fully grown
Water · 2,760 gallons
Energy · 3 million BTUs
Solid Waste · 185 pounds
Greenhouse Gases · 509 pounds

CALCULATIONS BASED ON RESEARCH BY ENVIRONMENTAL DEFENSE AND THE PAPER TASK FORCE. MANUFACTURED AT FRIESENS CORPORATION.